Estimating for Builders and Surveyors

Second edition

Ross D. Buchan, F.W. Eric Fleming and Fiona E.K. Grant

BUTTERWORTH HEINEMANN

OXFORD AMSTERDAM BOSTON LONDON NEW YORK PARIS
SAN DIEGO SAN FRANCISCO SINGAPORE SYDNEY TOKYO

Butterworth-Heinemann
An imprint of Elsevier Science Limited
Linacre House, Jordan Hill, Oxford OX2 8DP
200 Wheeler Road, Burlington, MA 01803

First published as Estimating for Builders and Quantity Surveyors 1991
Reprinted 1991
Revised and reprinted 1993
Reprinted 1995, 1996, 1998, 1999
Second edition 2003

British Library Cataloguing in Publication Data
A catalogue record for this book is available from the British Library

ISBN 0 7506 4271 8

Typeset in 11/12 Times by Integra Software Services Pvt. Ltd, Pondicherry, India
Printed and bound in Great Britain

Contents

Author's note

Repeated use of 'he or she' can be cumbersome in continuous text. For simplicity, therefore, only the male pronoun is used throughout the book. No bias is intended as the position of a foreman or craftsman can equally apply to a female worker.

Preface

This book has been written as an introduction to estimating for students taking courses in building and surveying at colleges, polytechnics and universities up to BSc level.

The content is devoted to the calculation of rates for the items in bills of quantities. All examples are based on items measured in accordance with the seventh edition of the *Standard Method of Measurement of Building Works* (SMM7). The chapters generally cover one topic or one trade; reference to the relevant sections of SMM7 is made in the chapter headings. The examples have been chosen to reflect both traditional and up-to-date technology.

The examples throughout have been prepared using a spreadsheet program which has resulted in the rounding-off of subtotals and totals, so that in one or two instances there would appear to be an error of a penny or two. This does not affect the final rate calculated, as the computer always works to a minimum of eight places of decimals (although perhaps only displaying or printing the result to two places).

The policy for all the trades included has been to provide a core of the most useful examples and to reinforce areas which appear to have been less well explored by other authors by giving rather fuller explanation. Feedback, via the publishers, on areas which the reader may feel require more detailed explanation, would be most welcome.

Finally, we have had a good deal of help from many colleagues in the construction industry, too many, in fact, to name individually here. To all of them we are most grateful and trust that a general acknowledgement of that help here will be acceptable.

R.D. Buchan
F.W. Eric Fleming
F.E.K. Grant

Note

The cost of labour in the examples is given as:

- Tradesman: £8.00 per hour
- Labourer: £6.50 per hour

Both rates are inclusive of tool and clothing allowances. However, where a labourer is paid an additional sum for continuous skill or responsibility, this sum has been added net to the rate (see the examples in Chapter 2).

Extracts from British Standards are included by permission of the British Standards Institution. Complete copies of British Standards may be obtained from the BSI by viewing their web page, http://bsi.org.uk

With Edition 2 a disk of examples is not supplied; however, worked examples from this book can be downloaded from the publisher's own website. The address is www.bh.com

1
Introduction

The UK construction industry

The UK construction industry employs one in ten of the UK working population, containing approximately 200 000 contracting firms with half of this figure being composed of private individuals and one-person firms. During 1998 the value of output of the whole industry was £56.3 billion, representing 9 per cent of the GDP (Statistics, http://www.ciboard.org.uk/Stats/stats.htm, accessed 15/01/02).

From a report produced by the Department of Trade and Industry (The State of Construction Industry Report, Issue 12, published Autumn 2000, http://www.dti.gov.uk/construction/stat/soi/soi12.htm, accessed 15/01/02), it has been stated that the balance between new and repair and maintenance work reached a plateau after a period of domination by new work. The impact of the Government's Spending Review 2000 (SR2000) allocates major investment to the areas of infrastructure and repair and maintenance, with housing and regeneration spending planned to increase by approximately 16 per cent per annum, between 2000 and 2001 and between 2003 and 2004. The Budget report of autumn 2000 introduced a number of policies to the significant benefit of construction work through a mixture of VAT incentives and stamp duty measures (The State of the Construction Industry Report, Issue 12, Department of Trade and Industry).

Rethinking Construction

The client's, and very often the public's, perception of the construction industry is one of poor management, poor quality, and poor value of products leading to poor service and resulting in premature repair and/or replacement. This negative image led to the UK government commissioning a report, which was completed by Sir John Egan, Chairman of the Construction Task Force, called 'Rethinking Construction' published in 1998. The key objectives of the report are to improve the industry's performance through the application by clients and industry of best practice.

The principal areas recommended in the report include targets to be achieved in productivity, profits, defects and reduced accidents. To achieve these targets four strands (*What is Rethinking Construction*?, http://www.rethinkingconstruction.org/about/intro.htm, accessed 08/01/02) have been initiated and Demonstration Projects will be adopted, which exemplify some of the innovations advocated in Sir Egan's report.

The four strands are:

1. The Movement for Innovation (M4I) which has established key performance indicators (KPIs) as a means of measuring and comparing construction company performance with the rest of the industry, and to assess improvement in the industry as a whole. The KPIs are divided into two sections, one to measure project performance, which includes client satisfaction with the product and the construction process, defects, predictability of cost and time, actual cost and time. The second section addresses company performance and measures profitability, productivity and safety.

2. The Housing Forum, which encompasses the private, public and social sectors, covers refurbishment, repairs and maintenance and new build. Its main goal is to gather together the various parties involved in the house building supply chain who are committed to change, innovation and renovation in construction. Using Demonstration Projects, experience in achieving quality, efficiency, sustainability and value for money in all sectors can be explored and developed with all the parties involved, i.e. house builders, professional institutions, landlords, local authorities, contractors, consultants and end users.
3. The Local Government Task Force (LGTF) was launched in March 2000 to advise local authorities on implementing the Rethinking Construction agenda. Various working groups were formed to address the issues and since their launch a number of publications have been produced. These include Repairs and Maintenance in Social Housing, The Best Value and Partnership Approaches to Procurement and the Rethinking Construction Implementation Toolkit which was launched in May 2001. The current forum of working groups is: Constructiononline, Building Control and Planning, Funding Streams and Wellbuilt.
4. The Central Government Task Force (CGTF) is chaired by HM Treasury and includes representatives from major Central Government Client Organizations, its role is to act as a pathfinder and head initiatives on best practice for government clients. The CGTF also provides information for the Office of Government Commerce, which has provided a series of guidelines to advise government clients on best practice. The guides hope to simplify, improve and drive forward best practice by improving central government procurement.

Sustainability

The construction industry is very limited in contributing to sustainability of the environment and there are potential areas where improvement could be made. In 1999 the Government published 'A better quality of life – a strategy for sustainable development in the United Kingdom' which set out a strategy to achieve an increased quality of life.

It was recognized that the construction industry has an essential and important role in addressing these issues through improving the quality of life, by providing safe, secure buildings for people to live and work in, and by ensuring that the industry itself works in a sustainable way, conserving resources, reducing pollution and waste and valuing its workforce (Sustainable Construction, http://www.dti.gov.uk/construction/sustain/index.htm, accessed 10/01/02).

A publication by the Government was produced following consultation with the industry in April 2000, 'Building a better quality of life – a strategy for more sustainable construction', with a review on progress being produced in October 2001. It was reported that progress was being achieved and measured using KPIs, although in some areas research was required to provide the tools and guidance on sustainable building practice and information on design advice.

Introduction to estimating

Regardless of size, shape or function of any proposed work, a method of producing an accurate estimate of the project is essential. A contractor cannot afford to make an unrealistic offer to a client who is instructing the work being undertaken. Estimating is the process of pricing work based on the information/specification and/or drawings available in preparation of submitting an offer to carry out the work for a specified sum of money.

This specified sum is known as the 'tender sum' and will be made in the context of a form of contract, which will include the condition under which the specified sum may be varied. The estimator must be confident that the tender sum is accurate as the standard form of building contracts does not permit the contractor to recover additional sums of money from the client due to errors in the estimating process.

This book describes the methods used by an estimator to compile prices or rates for items described in *Standard Method of Measurement*, 7th Edition (SMM7) format. Also included are chapters covering computer-aided estimating, tender strategy, cash flow and prefabricated structures.

Each chapter is a self-contained unit and, notwithstanding the fact that prices in the chapters interrelate, chapters can be read singly and in any order.

The invitation to tender

The method by which the client invites a contractor to tender can arise in various ways depending upon how simple or complex the work is and its value.

Verbal or written outline description

For straightforward low-value work, i.e. electrical wiring of a small house extension or, perhaps on a larger scale the external decorating of windows for an office, the client will generally invite the contractor to view the work, detail the specification/requirements to be carried out, and wait for the contractor to submit an offer.

It is then the responsibility of the contractor's estimator to measure the work, price it and make an offer which is based on the contractors' own terms and conditions. The offer may also contain details of specification but the contractor, in this situation, will be bound to carry out the work based upon the normal trade practice of the time.

Drawings, specifications and a form of contract

As work becomes more complex and detailed (for example, a family home or small factory unit) an architect will traditionally be appointed by the client, who will be instructed to design and complete drawings along with a specification and details of the standard form of contract. This information will be submitted to a contractor who will measure the quantities of work, price it and make an offer based upon the information received from the architect.

Complete design and build service

Alternatively, the client can approach a contractor and invite an offer for the complete works including the design and construction. Usually reserved for the type of work described in the previous paragraph, the design and build method of tendering is becoming more common and the scale of projects is becoming more complex. A procedure for the selection of a design/build contractor is described later in this chapter.

Drawings, bill of quantities and a form of contract

Medium and large-scale projects require a structured procedure to monitor and control the various aspects and individuals involved in the construction of projects. The main role of leading the design team and co-ordinating the various professionals involved has traditionally been the architect's, although more recently other construction professionals are taking on this responsibility. The design team leader is approached by a client with a list of requirements, known as the 'client's brief'; this information is used to produce the initial sketch designs. This preliminary design work allows the quantity surveyor to arrive at a cost estimate or tender value based on the client's brief. Depending upon how specific the information obtained from the client is and the detail of the sketch design, the initial tender value may differ considerably from the final account due to the minimal detail available at the initial stages.

If the predicted tender value is acceptable to the client, the architect will proceed with the detailed design in conjunction with structural, mechanical and electrical engineers and other consultants. The detailed design is given to the quantity surveyor who will prepare a list of all works involved called the 'bill of quantities', which contains the specification and the measured items of work calculated in accordance with a set of rules for measurement. He will also prepare a breakdown of the costs.

The set of rules for measurement is compiled by agreement between the Royal Institution of Chartered Surveyors (RICS) and Construction Confederation (CC). This set of rules, termed the *Standard Method of Measurement for Building Works*, was first compiled in 1922 and is currently in its 7th edition (SMM7), which was published in 1998. Scottish practice used the Scottish Mode of Measurement until 1963, when SMM5 was adopted for use in Scotland. Since then, both SMM6 and currently SMM7 have been accepted for use in Scotland.

SMM7 enables a project to be broken down into measurable work items. It contains rules of inclusions and exclusions of certain elements, which are known as coverage rules. It is important for the contractor's estimator to be familiar with these rules as under- or overpricing may result if it is thought an item of work is deemed included or not. For example, in the groundwork rules, D20.6, 'Working space allowance to excavations' is measured in m^2 and is deemed to include the cost of work associated with 'Additional earthwork support, disposal, backfilling, work below ground water level and breaking out' (C2). An estimator who does not realize the extent of the 'deemed to be included' items might underprice this item.

Once completed by the quantity surveyor, the bill of quantities is sent with the drawings to the contractor. It is the function of the contractor's estimator to price the items of work measured in accordance with SMM7, allowing for items with cost implications due to the coverage rules as 'deemed to be included'.

Due to the nature, value and complexity of a project, the invitation to tender will be expressed in one of the four ways described above.

The estimating method

On receipt of the instructions, specification, drawings, bill of quantities and standard form of contract from the client's representative, the contractor's estimator's task is to price the project and produce an estimate. Initially the estimator will subdivide the work into individual sections to be priced. With a bill of quantities the work will have been submitted to the estimator in sections, if not, the estimator must prepare these sections for pricing.

Sections or packages of the project, which are to be carried out by subcontractors, will be identified by the contractor's estimator. To ensure competition several subcontractors will be invited to tender for each of the work packages and will send to each a copy of the relevant section of the bill of quantities. The successful tenderer's prices will, with an addition for the contractor's overheads and profit, be incorporated into the contractor's offer.

The remainder of the work will be priced by the estimator based on sources of data from previous jobs, rates and experience. In each work section rates for items of work will be built up by calculations (in the manner which is described in the chapters of this book) based on the key resources required, mainly, labour, plant, materials plus an allowance for overheads and profit.

Labour

The estimator will have to price the time it will take for an operative or a gang to complete an item of work; a Table of Labour constants will be available, which are normally expressed as an output of work per hour. By totalling the hours and multiplying by a cost of labour per hour the estimator can determine the labour cost associated with the work item.

Plant

Generally, plant is priced on the same basis as labour, if the plant is used for a specific task. For example, an excavator digging a drain trench will be priced on the basis of the amount of work that it can do in one hour. Static plant, which cannot be associated solely to one item of work, such as scaffolding, site hutting or even a tower crane will be priced on the basis of the time that it is required on site. The cost is added to the tender as a lump sum and is not priced per unit of task, e.g. £1.50 per m^3.

Materials

The price for material is based on its purchase price with allowances for delivery, off-loading, storage and placing in position. In addition, an allowance must be made to cover wastage which may arise as follows:

- Handling/breakages – Brittle materials such as bricks, clay tiles and precast concrete paving slabs, etc. have a high breakage rate even allowing for the advances in mechanized handling of material which is wrapped and delivered to site on pallets.
- Site losses – An allowance has to be made for loose material such as sand and aggregate which, when delivered and tipped on site will, to some extent, be trampled into the ground, washed away by rain, etc.
- Cutting losses – Sheet materials are manufactured in standard sizes. Therefore plaster-board, plywood, glass, carpet, etc. must be cut to fit. Where the material is patterned or has to be cut to a difficult shape, losses can be considerable.

The CDM Regulations

The CDM Regulations 1994 came into force on 31 March 1995 and apply to virtually all construction works from concept designs through to project completion and beyond. They affect everyone who takes part in the construction process – the client, designers and contractors.

The Regulations reflect a fundamental shift for health and safety in the UK construction industry. They impose a management structure for construction sites. Imposition of this structure was deemed essential to reinforce the fact that:

- clients are in a position to exercise control over the level of safety adopted by the contractors offering bids;

- designers should be held accountable for the safety of their designs, in much the same way as equipment and machinery designers; and
- contractors should consider safety at the tendering stage and at all stages of the construction project.

The Regulations are also distinct from other health and safety regulations due to their breadth of application. The CDM Regulations apply to virtually all construction work during five distinct stages:

1. Concept and Feasibility
2. Design and Planning
3. Tender/Selection
4. Construction and
5. Commissioning and Handover.

The regulations apply where construction work is to take over 30 days or involve more than 500 person days of construction work or where the work involves five or more persons, but in this situation the work will not be notifiable to the HSE unless the 30 day rule is exceeded. Notification is not necessary, nor do the Regulations apply to work which involves four persons or less on site and takes less than 30 days.

The CDM Regulations identify five key 'posts' which have responsibilities for ensuring health and safety matters are addressed throughout the whole construction project. They are:

1. The Client or Client's Agent
2. The Designer
3. The Planning Supervisor
4. The Principal Contractor
5. Other Contractors.

Overhead and profit

In order to remain in business every contractor needs to cover their overheads and produce a profit for the future viability and expansion of the firm. Overheads and profit will be added to the cost of all work items to cover the cost of running the contractor's establishment and to provide a profit. The establishment cost includes the head office running costs for the various departments, such as estimating, buying, planning and accounts, together with those departments that are essential to any commercial firm such as marketing, strategic planning, public relations, personnel and training.

The amount added for overheads and profit is normally given as a percentage, the allowance generally being decided by the contractor's board of directors. The estimator will provide the cost for labour, plant and materials, leaving the deciding factor in a competitive tender of whether the contract will be won or lost dependent upon the percentage addition for overheads and profit. The contractor has four choices for the distribution of overheads and profit:

1. *Percentage addition to all individual rates.* This has an advantage where a variation to the work adds additional quantity, since the contractor will recover additional overheads and profit, but the converse is also true.
2. *Add the total amount to one item in the preliminaries section.* Usually to the supervision item. This has the advantage that the amount cannot be reduced, and puts the contractor

in a strong negotiating position where a claim is made for additional supervision owing to the contract period being lengthened.
3. *Addition to the summary total at the end of the bills of quantities.* This reveals the overheads and profit addition, which can prove to be a disadvantage if an overestimate of provisional sums and provisional quantities is made in the bills of quantities. A remeasurement of these items may lead to a reduction in the contract sum and hence also in the blanket addition.
4. *Add the amount selectively to specific items in the bills of quantities.* This will improve cash flow. This has advantages, but is also dangerous where work is subject to remeasurement.

The reasons considered by the contractor for the addition of overheads and profit are discussed at length in Chapter 22 on Tender strategy. In this book a 20 per cent addition is used for the purposes of illustration.

Tender procedure

The method employed to select a contractor to carry out work can be done in one of three ways:

1. Open tendering
2. Selective tendering
3. Negotiated tendering.

Open tendering

Details of the proposed project are advertised in the local or trade publications. The advert would include:

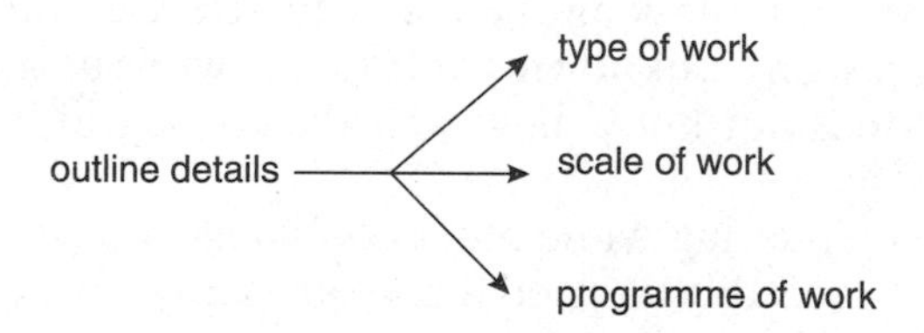

Contractors wishing to tender reply to the advertisement and receive the contract documents from the design team leader or architect. In some instances contractors may be required to pay a deposit. This covers the cost of documentation and discourages idle curiosity. The deposit is refunded upon receipt of a bona fide tender. Any contractor, irrespective of size or capability, may apply for the documents and tender. The advertisement is not legally binding and the employer (client) is not bound to accept the lowest tender.

Once completed all tenders will be returned by the stipulated date and time. It is then the decision of the client with advice from the design team to choose the successful contractor. Local authorities generally use open tendering, but its use has been declining in recent years.

Advantages of open tendering

1. Unknown contractor can tender for the work.
2. There is no restrictive list of tenderers, which does not allow favouritism – a valid point for local authorities who are publicly accountable.
3. There is no obligation to tender therefore all tenders received will be genuine.
4. Open tendering secures maximum competition.

Disadvantages of open tendering

1. Cost of tendering is expensive to the client who must bear the cost of reproducing multiple copies of drawings, bills of quantities, etc.
2. It is a lengthy operation requiring skilled estimating, the cost of which must be recovered on the job by the contractors. The higher the proportion of unsuccessful tenders the higher the cost to be recovered on the job.
3. The wrong contractor can be chosen. Little may known about the contractors – their record, experience, standard of workmanship, etc.
4. The lowest tender may not necessarily be a 'bargain'. Choosing a low tender may result in:
   - POOR WORK – a large number of, or even permanent, defects may occur unless there is close supervision by the client's agent.
   - POOR ORGANIZATION – late completion, specialist subcontractors delayed, etc.
5. A contractor may be awarded work for which he has little or no experience and which he may be ill-equipped to deal with.
6. A contractor who has under-priced his tender to win the contract may try to recoup this shortfall through claims or hard and unsupported bargaining on the final account.
7. The worst scenario is that the contractor may become insolvent due to incompetence or low pricing on jobs. This will involve the client in expense and delay.

In summary, open tendering is a method which allows new or unknown contractors to break into the market, while maximizing competitiveness, and allowing the client to select a contractor offering one of the lowest prices. However, using unknown contractors without an established record is a risky option, as the client does not know how well the work will be carried out.

Many good construction contractors avoid open tendering while others only resort to it if they are desperately struggling to win work. This is probably as good a reason as any for the client to consider alternative tendering options.

Selective tendering

This is the traditional method of awarding construction contracts. A number of contractors of known reputation are selected by the design team to submit a price for the project. The contractor who submits the lowest tender is generally then awarded the contract.

Tender list

The list of contractors will be made early in a project to allow the contractor to tender for the project if they wish. This will range from five to eight contractors depending on the type of work and value of the project.

The initial enquiry will sufficiently detail the project for the contractors to obtain a realistic appraisal of what is involved and to indicate whether or not they are interested in obtaining documentation (detail of information as advert for Open Tendering).

Inclusion on the list of tenderers will depend upon the character of the project and the size, i.e. a contractor who builds motorways will not be geared to undertake an alterations project. Local authorities and government agencies keep a general list which contractors can apply to join. Contractors are carefully vetted before being put on the general list. From the general list, a 'tender list' is made for specific projects.

Areas to consider while compiling a general or tender list would include:

1. Standard of workmanship;
2. Size of company;
3. Contractors' practice to sub-let work;
4. Reputation to meet completion dates; supervise quality of work and settle final account;
5. Examination of Boards of Directors – 'a sine qua non';
6. Financial stability – length of time in business, financial checks, bank references, etc;
7. Capacity available in relation to current workload;
8. Labour relations, number of disputes and stoppages in recent jobs;
9. The company's real willingness to tender.

It may be necessary to visit a recently completed project or one still in progress by a contractor on the selected tender list to see how they operate on site.

Once a general list is established it can be categorized into areas of work, i.e. road construction, house contractors, refurbishment contractors, etc. The established lists can be used on future projects but it is important that they are updated on a regular basis. If a formal list is not maintained an ad hoc tender list can be drawn up of contractors who are known to be suitable for a particular project. Alternatively, contractors can be invited by an advertisement, as in open tendering, to submit their name for inclusion in the list for a particular project – only those genuinely interested would reply.

The *Code of Procedure for Single Stage Selective Tendering* was produced by the National Joint Consultative Committee for Building (NJCC) and published by RIBA Publications in a revised format in 1989 and again in 1996. It makes the recommendation that no more than six contractors should be selected for any tender. Many commentators regard the maximum number stated as too high, bearing in mind the high cost of tendering.

Negotiated tender

Under this method only one contractor is approached, normally because the skills of the contractor are such that the architect and the other members of the design team wish to take advantage of the contractor's specialist knowledge at the design stage.

Following the completion of the design the contractor will price the bill of quantities and then enter into a negotiation with the quantity surveyor. For this type of situation the NJCC has issued the *Code of Procedure for Two Stage Selective Tendering*, which is discussed later.

Selective tendering provides a restricted but adequate list of technically suitable contractors of comparable standing, capable of carrying projects through in a reliable manner.

Selective tendering can be divided into two categories:

1. Single stage selective tendering;
2. Two stage selective tendering.

Single stage selective tendering

This procedure is given in the NJCC – *Code of Procedure for Single Stage Selective Tendering*. It recommends that a list of tenders is selected as previously described: the employer, architect, quantity surveyor and engineer would give input into the selection. Normally six to eight contractors are chosen with two in reserve. The contractors are contacted in writing to enquire if they would be willing to tender (see Appendix A of the *Code of Procedure for Single Stage Selective Tendering*). The enquiry should indicate that should the contractor decline to accept this invitation it will not prejudice a future invitation. This is important since many contractors think it is better to submit a 'high' cover price than decline an invitation.

On the date stated in the preliminary enquiry the documents forming the tender pack are dispatched to those contractors who have agreed to tender and have been selected for the final tender list. The tender documents include:

(a) two copies of the Bill of Quantities (one bound plus one unbound);
(b) two copies of general arrangement drawings;
(c) two copies of form of tender (see Appendix C, of the *Code of Procedure for Single Stage Selective Tendering*), (Figure 1.1);

Tender for: (description of the works)

To: (Client)

Sirs,
We having read the conditions of contract and bills of quantities delivered to us and having examined the drawings referred to therein do hereby offer to execute and complete in accordance with the conditions of contract the whole of the works described for the sum of:

£...........................

and within .... weeks from the date of possession.

We agree that should obvious errors in arithmetic be discovered before acceptance of this offer in the priced bills of quantities submitted by us these errors will be dealt with in accordance with Alternative 2 contained in Section 6 of the Code of Procedure for Single Stage Selective Tendering.

This tender remains open for acceptance for .... weeks from the date fixed for the submission or lodgement of tenders*.

Dated this .... day of .... 20 ....

Name:

Address:

Signature:

Witness 1:

Witness 2:

* Scotland only: 'unless previously withdrawn'.

Figure 1.1 *Tender form*

(d) instructions for the return of the documents, e.g. an envelope which states the date, time and place of return, marked with a code which indicates to the quantity surveyor the name of the tenderer (see Appendix B of the *Code of Procedure for Single Stage Selective Tendering*); and
(e) invitation to tender letter (see Appendix B of the *Code of Procedure for Single Stage Selective Tendering*).

A standard letter should go to those who indicated their willingness to tender but have not been selected for the final list, stating that they will be invited to tender for future work.

Decision to tender

Having received the invitation the contractor must make a decision as to whether to tender or not. This will be dependent upon:

- The strategic management policies of the company, based on the plan and the determination of whether the likely cash flow demand of the project is conducive to the company cash flow and borrowing arrangements. (These factors are discussed at length in Chapter 22 on tender strategy.)
- The resource demand of the proposed project in comparison with the resources, which the company has available, and also whether these resources can easily be deployed in the geographic area of the proposed site.
- The type and form of contract. Currently there are many ways of procuring buildings and the contractor must consider whether it is desirable to operate under a form of contractual arrangement, which is not familiar. The contractor must also ascertain the extent of risk associated with a particular project and decide whether the risk is worth taking.
- The extent of the competition, since this will affect the probability of winning the tender; for example, a contractor may make a decision not to tender in a competition with more than four others. Although not stated in the *Code of Procedure for Single Stage Selective Tendering*, the design team should indicate in the preliminary enquiry the target number of contractors to be included in the tender list.

Pricing the tender

To price the tender the estimator will firstly identify the areas of the work which will be undertaken in house and those which will be subcontracted out. The estimator will pass a copy of the bill of quantities to the buying department so that quotes for materials may be obtained. It is at this point that most contractors discover who the competition are, simply by selecting the easiest defined or least common material in the bill of quantities and asking the supplier if any other contractor has requested an identical quantity of the same material. It is very difficult to keep a tender list confidential to the design team.

Depending on the complexity of the project and the required sourcing of materials by the contractor or any of the subcontractors to comply with the performance specification defined in the bill of quantities, the design team should plan for at least six weeks before the return of a priced tender. Tradition states four weeks for any tender, but this must be extended if the contractor is expected to carry out any design work.

The estimator will generally price all items in the bill of quantities at cost with no mark-up for overheads and profit. Following this exercise the management of the contractor will decide upon the overheads and profit percentage mark-up. They will consider the items of risk in the project allowing for this within the percentage and will adjust the projected cash flow by selective addition of the overall amount of overheads and profit (see Chapter 22 on tender strategy).

If a particular risk is so great that the contractor is unable to account for it within a realistic tender, the tender document may be qualified. For example, an item in a bill of quantities may state (contrary to SMM7): '*17C Carry out all necessary repairs to the Welsh slate roof ------- item.*' In this situation a lump sum is being requested. The contractor may consider the risk of pricing this item as a lump sum to be too great, and therefore inserts in the tender document: '*We offer to carry out the work . . . for the lump sum of £300 000, which includes a provisional sum of £1000 for repairs to the roof in place of bill of quantities item 17C.*' Although such a statement is a legally valid offer under the law of contract, the Code states that in such circumstances the tenderer should be given the opportunity of withdrawing the qualification. If the contractor is unwilling to withdraw the qualification, the tender should be rejected.

The logic behind this argument is that tenderers have the opportunity to bring such matters to the attention of the design team during the tender period, and a correction, where this is thought appropriate, can be issued to all tenderers. In practice many tenders received are qualified, and the design team are put in a difficult position when under pressure from the client to accept the lowest.

Receipt of tenders

Two days before the date for the receipt of tenders the quantity surveyor should determine which tenders have not been received by reference to the coded envelopes. The contractors whose tenders have not been received should be telephoned to ascertain whether a tender is to be submitted. In the case where tenders have been delayed in the post this permits another copy to be sent by post, courier or fax.

As soon as possible after the time for the receipt, all tenders should be opened. In Scotland the priced bills of quantities will have been sent with the tender in a separate envelope and the bills of the lowest tenderer will be opened for checking. Elsewhere in the UK the priced bills will be requested from the lowest tenderer.

In the event that no arithmetical errors are found and the offer is acceptable to the client, the offer will be formally accepted and a contract entered into. The bills of the remaining tenderers in Scotland will be returned unopened.

The Code of Procedure recommends that immediately after the opening of tenders all but the three lowest should be informed that their tender was unsuccessful. If an arithmetical error is found during the checking of the bills of quantities then two alternatives are given in the Code.

1. *Alternative 1* states that the tenderer should be notified of the error and given the option of confirming the tender or withdrawing.
2. *Alternative 2* states that the tenderer should be notified of the error and given the option of confirming the tender or correcting the error. In reality the difficult decision only applies where the correction of the tender puts the tender price higher than that originally. In the case that the tenderer withdraws (Alternative 1) or that the corrected tender is higher than the second lowest (Alternative 2), the priced bills of quantities of the second lowest are examined and the whole procedure starts again.

Feedback

It is recommended that each tenderer is given a list of the tender sums received so that each can observe their position in the list and a decision can be made with regard to the strategy to be adopted for future tenders. In Scotland, as a separate list to the tenders, each tenderer will receive an alphabetical list of the contractors tendering.

Serial tendering

A variation on the above procedure is that of serial tendering in which the contractor tenders for a series of similar contracts. This method was popular during the period of industrialized building, particularly in conjunction with school buildings, where a learning curve was anticipated. The rationale was that the contractor might accept a small loss on the first school building, but by the time the third had been constructed the contractor's operatives would be fully conversant with the system. Serial tendering also guarantees work for an extended period. The method is based on the tenderer pricing a master bill. This will contain the quantities for one school but may include, for instance, three options for the construction of a roof, so that three sets of roofs for one school will appear in the master bill. Similarly the foundations may be measured as both pads and piles (assuming a light steel frame construction).

The tender procedure will be exactly as the above but the contractor will know that the pricing of the master bill effectively represents a tender for five schools to be constructed over the following five years. The schools are designed and bills of quantities prepared and priced at the rates in the master bill. Normally the serial contract is based upon the rates being adjusted in accordance with a price index, for example the national economic development office (NEDO) index.

Two stage selective tendering

The majority of negotiated tenders, as stated above, arise from a desire to involve the contractor in the design phase of a project. This method is used when it is necessary to appoint a contractor at an early stage, i.e. before the scheme has been designed. It can be described as a number of procedures that are suited to contracts where:

- Time is of the essence;
- The contract is so complex as to warrant the appointment of a contractor at the earliest possible stage.

Two stage tendering is fully explained in the NJCC *Code of Procedure for Two Stage Selective Tendering*. It describes a method which allows the negotiation of a tender with an element of competition, a brief outline is given below.

Stage one

1. Preliminary enquiry sent to short list of contractors.
2. First stage documentation prepared based on preliminary information from architect, engineers, etc. usually an Approximate B of Q (the approximate bill of quantities is a document containing an outline specification and provisional item descriptions and quantities).
3. Tender documentation is sent to the contractors in the same way as for Single Stage Selective Tendering.
4. The contractor is selected and the first stage of the tender is accepted.

The Code states that tenderers should be given five weeks to prepare their tender. It recommends that the tender is submitted under the same rules as for single stage selective tendering, with the same procedures for notification, errors, etc.

However, many design teams do not use the preliminary tender based upon the notional bill as the sole method of selection. Instead they will invite each tendering contractor to a meeting to discuss the contractor's approach to the project. The approach will be discussed in terms of the contractor's contribution to the design, the previous experience of such schemes, the personnel to work with the design team and their previous experience. The aim is to ascertain whether the contractor and the design team can work well together. It is the combination of these meetings and the preliminary tenders which determines the successful contractor.

The acceptance should be clearly defined in respect of:

(a) either party being entitled to withdraw from the second stage;
(b) entitlement and method of ascertaining costs which may be incurred by the employer and/or contractor in the second stage should the parties fail to proceed; and
(c) method of reimbursement agreed for any work done on site before acceptance of the second stage.

Stage two

1. Detailed design carried out by architect and engineer.
2. Contractor is appointed to advise design team.
3. Preliminary work is carried out on site, if possible.
4. Quantity surveyor prepares second stage documentation, firm bill of quantities. Enters into negotiation with contractor and arrives at contract price.
5. Agreement is reached, contract signed and major works started on site.

An agreement will be made with the successful contractor regarding the cost of the contractor's input before the commencement of the design stage. The design stage is undertaken. Bills of quantities are prepared by the quantity surveyor and priced where possible at the rates in the approximate bill. Those items, which cannot be priced on this basis, are priced by the contractor and are subject to negotiation.

Finally, following successful negotiations the contractor will make a formal offer to carry out the work for the sum of money negotiated, which will be formally accepted by the client. A standard form of contract is completed.

This method of tendering allows the client the choice of a contractor who is capable and keen to contribute to the design while retaining the important element of competition.

Two stage selective tendering allows the contractor to be appointed at an early stage in the project. This means that a number of different professions must work together as a team. Due to the conservative nature of the construction industry, the fragmentation into specific specialisms and the guarded nature of professionals for their own professions, teamwork is sometimes difficult to achieve. Survival in today's world has forced these dilemmas to be set aside and eliminated. A construction professional puts the needs of the client first to win business and their place in the market.

Selective tendering for design and build

In October 1985, the NJCC published a *Code of Procedure for Selective Tendering for Design and Build.* Similar to the two stage method described above, this outlines a tender method which is considered fair to all parties. In design/build particularly, a tendering contractor is only prepared to pay for the development of the design and tender if there is a good probability of winning the work.

The Code's recommendations are as follows:

1. Six to twelve selected contractors are circulated with a letter of preliminary enquiry, which sets down the nature of the project, the anticipated start and completion dates, etc. and asks whether the contractor would be interested in tendering for the project.
2. Those contractors who reply positively to the preliminary enquiry are sent a copy of the employer's (client's) requirements. The documents which represent the employer's requirements may be as little as a description of the accommodation required, or may be anything up to a full scheme design undertaken by a full design team employed by the client. However the employer's requirements are expressed, it is anticipated that they will be fully considered at the tender stage and will not be subsequently changed.
3. Tendering contractors are called for an interview with the employer and his advisers. At the interview the following will be discussed:
   - The form of construction anticipated by the contractor;
   - The contractor's organization and anticipated programme of work;
   - The nature and extent of the design liability insurance.
4. Following the interview a shortlist not exceeding four contractors will be drawn up and formal tender documents will be dispatched. The Code describes a large number of variations in tender practice which are acceptable at this point. These vary from a straight single design and lump sum tender to multiple alternative submissions and two stage selection. In the latter case the contractor may submit alternative schemes and give rates for major items of work. Following an analysis of the schemes received, one contractor may be given the go-ahead to develop particular schemes further. The resulting tender may or may not be open to negotiation based upon the rates originally submitted.
5. The final scheme prepared by the contractor and submitted with the tender is termed the contractor's proposals. The contractor's proposals are an answer to the employer's requirements, and comprise:
   - plans, elevations, sections or typical details;
   - information about the structural design;
   - layout drawings showing the services to be included; and
   - specifications of workmanship and materials.
6. If the contractor's proposals and tender are acceptable to the employer or client, a contract is signed between the contractor and employer based upon the submitted data.
7. Finally, it is recommended that feedback be given to the unsuccessful contractors, though this is unlikely to be as useful as a simple list of tender prices received.

Under the design/build contract the estimator has responsibility for the measurement and pricing of the work since no bills of quantities are provided. Many design/build contractors will engage professional architects and quantity surveyors to carry out their traditional roles.

The design/build contract is a lump sum contract payable in stages or by valuation in accordance with the conditions. As there is no provision for a bill of quantities and no schedule of rates, the contract makes provision for a 'contract sum analysis' to permit the valuation of employer's changes and for formula price adjustment if this is included as a contract provision.

There is no provision in the contract for prime cost sums and these should not be included. There is, however, provision for the expenditure of provisional sums included by the employer, and the contract sum analysis should be sufficient to allow this.

2
The cost of labour

The cost of labour represents a high proportion of today's building costs. As a percentage of a bill rate (other than labour-only rates) the figure varies from about 30 to 75 per cent over the various trades. In the overall cost of the building, the cost of labour will account for something like 40 to 60 per cent.

The all-in hourly rate

The complete cost of labour is calculated as an hourly figure which is widely known as the 'all-in hourly rate'. This should not be confused with the term 'all-in rate', which is sometimes used instead of 'bill rate' to describe the price attached to an item in a bill of quantities and which includes labour, material, plant, profit and overhead costs.

There are basically two ways to calculate the all-in hourly rate. The first is to consider what a particular class of operative will cost the building employer for one year and to divide this by the total hours worked in that year. The second is to consider what the same group will cost for one week and to divide by the hours worked that week.

The former is widely used by various general sources of bill rates such as pricing books, computerized libraries of rates and some building magazines. The all-in hourly rate produced can only be a typical average rate, as the producer of the data has no contract in mind and therefore must assume how much, if any, overtime is being worked, the ratio of tradesmen to labourers in squads, the amount of supervision and whether this is done by working or non-productive foremen and so on. There is no notion of distance from site, so averages can only be given for things like travelling time, fares, lodgings, transport etc.

The all-in hourly rate based on one week's work is produced by estimators for particular circumstances. In this instance the estimator knows how large the contract will be, how far away it is and what work is required to be done, and therefore the resources necessary in terms of operatives, supervision, overtime, travelling etc. In other words a lot of averaging and guesswork becomes a matter of judgement based on the employer's records of previous contracts.

The reader is recommended to study the all-in hourly rates given in the building magazines and price books. Notice the comments regarding the allowances for overtime, bonuses etc.

The Working Rules

Part of the cost of labour is naturally the actual wages paid to the workers carrying out the productive work. The workers are generally referred to as operatives and we will return to the idea of productive work later in the chapter. The remainder of the cost of labour is made up of a number of payments which the builder must make and which are a direct result of employing operatives. These payments are not always made to the operative and are not always on their behalf.

The following is a classification of labour costs:

- Payments made as a result of negotiation between unions and employers, including hourly wage rates, holidays with pay and overtime rates.
- Expenses such as fares, meals and lodgings.
- Statutory payments such as national health insurance, employer's liability insurance, sick pay and severance pay.
- Payment made as a matter of good business management, including third party liability insurance and adequate supervision.

There are several discrete bodies in the construction industry which agree rates of pay and conditions for different types of operative. The organization whose rules are reviewed in this chapter is the National Joint Council for the Building Industry (NJCBI). There are separate organizations negotiating on behalf of tradesmen such as plumbers and electricians, as well as alternative organizations negotiating for bricklayers, carpenters etc.

The NJCBI comprises members from the following organizations:

Employers:	Construction Federation
	National Federation of Roof Contractors
	National Association of Shopfitters
Trade Unions:	Union of Construction, Allied Trades and Technicians
	Transport and General Workers Union
	Guaranteed Minimum Bonus (GMB)

Negotiated hourly rates and the rules of payment of a long list of benefits for the employee are given in the Working Rules, a document produced by the NJCBI. The rates and rules are published with regional amendments through the individual regional offices. In practice, the building employer will use the rules appropriate to the area in which the work is being carried out. The various regional editions of the rules are obtainable from the regional offices of the unions or employers' federations.

Not all building companies pay NJCBI or any other nationally negotiated rates of pay. Negotiated rules are simply standard forms of employment contract. A building contractor may employ labour on any mutually agreed basis as long as statutory requirements are fulfilled. However, no matter what the basis for the employment contract, much the same types of cost are involved. Therefore the job of the estimator will follow similar patterns to those described in the text and shown in the examples.

The examples given in this chapter are based on the Working Rules approved by the Scottish Regional Committee and the NJCBI, 1999 edition with amendments to 2000. As the Working Rules are under continuous review, the reader is reminded here that they should purchase a copy of the most recent edition and keep it up to date as amendments are published. Amendments are widely circulated in the building journals and the house journals of various professional bodies such as the Chartered Institute of Building (CIOB) and the Royal Institute of Chartered Surveyors (RICS).

Estimating the cost of labour

It is important that the estimator gets the cost of labour correct; it is a major component of practically every bill rate. Having said that, it is equally important not to be bogged down in a plethora of unwarranted detail in the calculation of that cost.

With the word 'detail' in mind, let us look first of all at the list of factors which could possibly be taken into consideration in calculating the cost of labour. The factors are listed in Table 2.1 and those used in the examples at the end of this chapter are marked with an asterisk.

Table 2.1 *Factors affecting the cost of labour*

Factor	Used in Examples	Costing Option Table 2.2	Working Rule (WR)
Earnings:			
Hourly rates of pay	*	1	1
Guaranteed minimum weekly earnings	*	1	17
Overtime	*	1	4
Rotary shift work	Not included		6
Night work	Not included		7
Continuous working	Not included		8
Tide working	Not included		9
Tunnel working	Not included		10
Distance to site: Daily Fares and Travel Allowances			
Extent of payment	*	1	5.1
Measurement of distance	*	2	5.2
Transport provided free by the Employer	*	2	5.3
Transfer during working day	Not included	2	5.4
Emergency work	Not included		5.5
Extra payments:			
Continuous skill or extra responsibility	*	1	1.2.2
Intermittent responsibility	*	1	1.4
Holidays:			
Holidays with pay scheme	*	2	18
Public holidays	3.43%	3	19
Retirement and death benefit schemes	*	2	21
Employers contributions for national insurance	Statutory amounts	3	
Injury and sickness:			
Injury payments	*	3 or 6	
Sick pay	£1.37/week		20
Payment for work in difficult conditions		1	1.4
Empoyer liability insurance	2%	3	
Construction Industry Training Board levy	*	1 or 3	
Redundancy funding	1.5%	3	
Supervision: foremen, chargehands, gangers	By calculation	2	
Rest and meal breaks	4.26%	3	3.1
Bonus, incentive, productivity schemes	Not included		2
Excess overtime allowance	*	3	

* Items so marked are the subject of rates or fixed amounts from the Working Rules and these have been applied in the examples.

Table 2.2 *Costing options*

1. Use the correct hourly rate, allowance or addition together with the correct hours.
2. Add the actual cost to the cost of labour.
3. Add an appropriate percentagc to the cost of labour.
4. Reduce the output per man by an appropriate amount.
5. Include the cost in the preliminaries.
6. Include the cost in overheads/oncost.

There are a number of ways in which these items can be costed; these costing options are listed in Table 2.2. For every factor listed in Table 2.1 there is a head office and site overhead/oncost for administering wages, holidays with pay, redundancy schemes, bonus, expenses etc. Costing option 6 in Table 2.2 should therefore be assumed to be in addition to those already listed.

In addition to the factors in Table 2.1 there are a great many items listed in the Working Rules on the subject of employer/employee legislation. These are not easily categorized as part of the cost of labour, but nonetheless they do cost the builder a considerable amount of money which must be recouped in the ordinary course of the business. The following list is by no means exhaustive:

- Liaison with safety representatives appointed by trade unions
- Time off for training safety representatives
- Safety committees with union and employer representation
- Time off and facilities for site, job or shop stewards to carry out their duties (the facilities are generally provided by the builder).

The costs of these are generally accounted for by an addition in the overheads/oncost (option 6 of Table 2.2).

It is interesting to note that options 4 and 5 in Table 2.2 are not thought suitable for any of the items listed in Table 2.1. For example, tea/meal breaks are shown in Table 2.1 as being most suitable for the use of costing option 3. Option 1 does not fit the circumstances, there being no hourly rate, allowance or addition. Option 2 assumes a fixed sum which again is inappropriate. Option 3 is the most feasible in this instance. The loss is easily assessed, occurs on every job and affects every employee. Option 4 would involve a very complicated calculation concerning the proportion of basic hours, productive hours, overtime hours etc., and the result would be no more accurate than that obtained with option 3. Options 5 and 6 are really non-starters, as tea breaks are hardly the kind of statistic one would expect a builder's costing department to keep. The estimator therefore gets no feedback on this cost in relation to turnover.

We now discuss the estimating of each of the factors in Table 2.1.

Entitlement to basic and additional rates of pay (WR1)

Hourly rates of pay

The idea of the basic week can be varied should the contract require shift work or night work, and then WR6 and WR7 are invoked.

Table 2.3 *Classifications of basic and additional pay rates of operatives*

Name	Job description	Weekly rates based on 39 hours (Current at June 2002)	Basic Pay (pence per hour)
General Operative	Carry out general building and civil engineering work	£214.11	549 p
Skilled Operative	Different levels depending on skill to be undertaken. Work levels are defined in Schedule 1, additional pay per hour is listed.	Skill Rate 4 £230.49 Skill Rate 3 £244.53 Skill Rate 2 £261.30 Skill Rate 1 £271.05	591 p 627 p 670 p 695 p
Craft Operatives	Carry out craft building and civil engineering work	£284.70	730 p

Operatives employed to work in the building and Civil Engineering Industry are entitled to basic pay in accordance with this Working Rule (WR 1). Classifications of basic and additional pay rates of operatives are shown in Table 2.3.

In addition there are the adverse condition classifications, A–E, listed in Schedule 2 which attract additional payments.

Of course a builder who is only willing to pay basic hourly rates either will not be able to attract employees or will attract only operatives whose performance is poor in quality and quantity or will have to have very good productivity deals or bonuses to compensate for low basic wages. Builders are forced by the shortage of skilled workers to pay in excess of these rates. The additional costs involved vary from region to region and from builder to builder, but are affected by the amount of work available, the number of builders bidding for that work and their individual success levels. The actual rates paid in practice may be augmented by a few pence, but could well be doubled where work is plentiful and labour short. The calculations used in this text will be based on the basic rates.

Work in difficult conditions (WR1 – Schedule 2)

The payments made under this working rule are added only to the basic hours the operative is employed under these conditions. The conditions are all fully described in WR1.4.1 and Schedule 2, as is the procedure to be followed should conditions occur which have not been foreseen when the rule was framed. The builder is expected to provide protective clothing, boots, gloves etc.

Bonus (WR2)

Guaranteed Minimum Bonus (GMB) was abolished in 1992, the money being added to the basic weekly wage then divided by 39 hours to give the hourly rate. The working rules state . . .

> *It shall be open to employers and employees on any job to agree a bonus scheme based on measured output and productivity for any operation on that particular job.*

Working hours (WR3)

Hourly rates of pay – basic working week 39 hours (Mon–Thurs 8 hours, Fri 7 hours). Basic week can be varied if contract requires i.e. shift work, night work, rotary shift work or tunnel work.

Rest/meal breaks – at each site there shall be a break or breaks for rest and/or refreshment at times to be fixed by the employer. The breaks, which shall not exceed one hour per day in aggregate, shall include a meal break of not less then half an hour.

WR 3.3 – no operatives shall undertake any jobbing, contracting or work of any kind after working hours, either on their own account or for a different employer.

Scottish amendment

The Scottish Regional Joint Committee of the Construction Industry Joint Council has decided to adopt the declaration made by the Scottish National Joint Council for the Building Industry on 3rd March 1947, with regard to tea breaks, namely

> *The Council agreed that the claim as tabled (that a forenoon break of ten minutes be allowed and paid for by the employer) should be withdrawn on a declaration by the Council that the taking of tea at the place of work is not considered a breach of the present rules.*

In our examples we will allow that 20 minutes are lost each day for morning and afternoon teabreaks. This amounts to a 4.26 per cent loss in production.

Overtime (WR4)

Once an operative has worked 39 hours, further working hours are designated as overtime.

Overtime would be calculated as follows:

Monday to Friday
For the first four hours after completion of the normal working hours of the day – time and a half; thereafter at the rate of double time, until starting time the following day.
Saturday
Time and a half, until completion of the first four hours and thereafter double time.
Sunday
At the rate of double time, until starting time on Monday morning.

Overtime is calculated on the normal hourly rates. Additional payment for skill and responsibility or adverse conditions and bonus are not included when calculating overtime payments.

Scottish amendment

When operatives start before the usual starting time they shall be paid at overtime rates for any time worked beyond the normal day. Starting time not to be before 6.00 a.m. and in cases where the operatives employed are unable to continue to work until the letting-off time provided by the working rule, the time worked before the usual time of starting shall be paid at time and a half.

The special cases of shift work and night work are dealt with in WR6 and WR7.

Once an operative has worked the normal 39 hours in a week, any further working hours are designated overtime. In general terms the overtime hours worked are multiplied by a factor and the basic hourly rate is then applied. This multiplier varies according to when the

Table 2.4 *Example of basic and overtime hours worked*

	Monday	Tuesday	Wednesday	Thursday	Friday	Total
Start	8.00	8.00	8.00	8.00	8.00	–
Lunch	0.50	0.50	0.50	0.50	0.50	–
Finish	18.00	18.00	18.00	18.00	18.00	–
Total hours worked	9.50	9.50	9.50	9.50	9.50	47.50
Basic hours	8.00	8.00	8.00	8.00	7.00	39.00
Overtime hours	1.50	1.50	1.50	1.50	2.50	8.50
Multiplier	1.50	1.50	1.50	1.50	1.50	–
Overtime hours paid	2.25	2.25	2.25	2.25	3.75	12.75
Total hours at basic rate	10.25	10.25	10.25	10.25	10.75	51.75
Productive hours	9.50	9.50	9.50	9.50	9.50	47.50
Non-productive hours	0.75	0.75	0.75	0.75	1.25	4.25

overtime hours are worked, which in turn hinges on the definition of the normal working day within which the basic hours are worked.

For example, a builder may wish the operatives to work from 8.00 a.m. to 6.00 p.m. on Monday to Friday. The normal starting time is 8.00 a.m. and the lunch break will be of half an hour duration. The operatives will be working the 39 hours basic time from 8.00 a.m. to 4.30 p.m. on Monday to Thursday and 8.00 a.m. to 3.30 p.m. on Friday. Overtime will therefore have to be paid for the periods in excess of these times, i.e. for 1.5 hours on Monday to Thursday and 2.5 hours on Friday. The reader should learn to tabulate such information; they will appreciate immediately that simply writing it out can lead to errors in examinations or in practice.

Table 2.4 shows the data for this example. The rows with start, lunch and finish times are obvious. The next row is the total hours worked: subtract start time from finish time less lunch break. (This is very easy with the 24 hour clock, which is used in the examples throughout this chapter.)

The basic hours come from WR3 and are subtracted from the total hours worked to give the overtime hours worked. The next row is the multiplier. To find the multiplier look at WR4.

Productive hours are the same as total hours worked. This total is divided into the total weekly cost (to the builder) of the operative to determine the true cost of each hour when output is achieved. Non-productive hours are hours paid less productive hours. This time is called 'non-productive' because the operative is paid for it but the builder receives no work in return.

Table 2.4 might look a little pedantic and the authors would not advocate its literal use. However, something of the kind can be of great help in setting the scene, not just for overtime but for other forms of time-based payment such as travelling time (see the section 'Daily Fare and Travel Allowance').

Daily fare and travel allowance (WR5)

This is no longer a simple repayment of expenses beyond set limits nor a payment based on a notional time travelled by the operatives rate.

Operatives are now entitled to a daily fare and travel allowance, measured one way from their home to job/site. Allowances are in accordance with published tables. There is no entitlement for

distances less than 15 km or in excess of 75 km. Employers may opt at their discretion to refund in full actual fares paid by the operative rather than pay the daily fare allowance.
Where the employer provides free transport the operative will not be entitled to the daily fare allowance.

Rotary shift working (WR6)

Rotary shift means a situation in which more than one shift of not less than eight hours is worked on a job in a 24 hour period and an operative employed on that job rotates between the different shifts (whether in the same or different pay weeks).

Night work (WR7)

Where work is carried out at night by a separate gang of operatives from those working during daytime, operatives shall be paid at the normal hourly rate plus an allowance of 25 per cent of the normal hourly rate.
Overtime shall be calculated on the normal hourly rate provided that in no case shall the total rate exceed double the normal hourly rate. Overtime shall be paid as follows:

Monday to Friday
After completion of the normal working hours at the rate of time and a half plus night work allowance (i.e. time and a half plus 25 per cent of normal hourly rate) for the next four hours and thereafter double time.
Weekends
All hours worked on Saturday and Sunday night at double time until the start of the working hours on Monday.

This rule does not apply to operatives on shift work, tunnel work or continuous working.

Continuous working (WR8)

This applies to operatives whose normal duties require them to be available for work during mealtimes and consequently have no regular mealtimes, who are also responsible for handing over to their counterpart at commencement and completion of duties. The operative will be paid for the number of hours on duty at the normal rate plus 20 per cent as a continuous working allowance.

Tide work (WR9)

Where work governed by the tidal conditions is carried out during part only of the normal working hour, and an operative is employed on other work for the remainder of the normal working hours, the normal hourly rate will be paid.
Where work governed by tidal conditions necessitates operatives turning out for each tide and they are not employed on other work, they shall be paid the minimum for each tide of six hours pay at ordinary rates provided they do not work more than eight hours in two tides.

Tunnel work (WR10)

The long-standing custom of the industry that tunnel work is normally carried out by day and by night is reaffirmed. Where shifts are being worked within and in accordance with the driving of tunnels the first period of the shift equivalent to the normal working hours specific in the Working Hours shall be deemed to be an ordinary working day. Thereafter the next four working hours shall be paid at time and a half and thereafter at double time provided that:

1. Work on a Saturday – the first four hours shall be paid at time and a half and thereafter at double time.
2. Shifts worked wholly on a Sunday – payment shall be at double time.
3. Shifts commencing on Saturday and continuing into Sunday – payment shall be made for all hours worked at double time.
4. Shifts commencing on Sunday and continuing into Monday – hours worked before midnight paid at double time and thereafter four working hours after midnight shall be calculated at time and a half and thereafter at double time.

Refuelling, servicing, maintenance and repairs (WR11)

An operative on mechanical plant shall, if required, work and be paid at their normal hourly rate for half an hour before and half an hour after the working hours prescribed by the employer for preparatory and finishing work of the equipment.

Refuelling, servicing, maintenance and repair work carried out on Saturday and Sundays shall be paid in accordance with the rule on Overtime Rates.

Storage of tools (WR12)

Where practicable the employer shall provide adequate lock-up or lock-up boxes where tools can be left at the owner's risk, provided always that the employer shall accept liability up to a maximum amount specified by the Council for any loss caused by fire or theft of tools properly secured. At all times operatives shall take good care of their tools and personal property and act in a responsible manner to ensure their reasonable safety.

Loss of clothing (WR13)

Where an operative leaves clothing in accommodation provided by the employer, the employer shall be liable up to a maximum amount specified by the Council for loss of such clothing through fire.

Transfer arrangements (WR14)

Rather than have operatives travel by public transport and risk delays etc., the employer may provide transport. In that case travelling time is still paid but obviously no expenses. The cost of this transport must be included in the cost of labour.

At any time during the period of employment, the operative may, at the discretion of the employer be transferred from one job to another.

The employer has the right to transfer an employee if the site is within daily travelling distance from the operative's home (i.e. within one hour of travelling when transport is provided by employer or within two hours using public transport). Transfer to a job that requires operatives to live away from home shall be by mutual consent. Consent is not required if the operative has received subsistence allowance within the preceding 12 months.

Subsistence allowance (WR15)

An operative necessarily living away from the place in which they normally reside shall be entitled to subsistence allowance of an amount specified by the Council.

Subsistence allowance shall not be paid if the operative is absent except through illness or injury. If the operative returns home due to illness or injury the employer will pay the operative's travel expenses. Operatives receiving the subsistence allowance will only be entitled to the daily fare allowance. In the worked examples the amounts for subsistence are deemed to have been agreed by the Council.

Periodic leave (WR16)

An operative necessarily living away from the place in which they normally reside shall be entitled to payment of fares or conveyance in transport provided by the employer as follows:

1. From a convenient centre to the job at commencement.
2. To the convenient centre and back to the job at the following periodic leave intervals:
   (a) For jobs up to 128 km from the convenient centre (measured in a straight line), every four weeks.
   (b) For jobs over 128 km from the convenient centre (measured in a straight line), at an interval fixed by mutual arrangement between the employer and the operative before going to the job.
3. From the job to the convenient centre.

Payment of fares

Where the employer does not exercise the option to provide free transport, the obligation to pay fares may, at the employer's option be discharged by the provision of a free railway or bus ticket or travel voucher, or the rail fare.

The convenient centre shall be a railway station, bus station or other similar suitable place in the area in which the operative normally resides.

Guaranteed minimum weekly earnings (WR17)

So far we have discussed rates of pay as if all operatives were to be fully employed. This is not always the case. A number of factors can cause work to cease, if not for the whole building site then for parts of it or for particular operations. Such factors include excessively wet or frosty weather delaying concreting or bricklaying; high winds preventing work on roofs and high scaffolds or forcing the shutdown of tower cranes; non-delivery of essential materials; and strikes by other operatives, suppliers or subcontractors.

Where these factors are not the fault of the operatives and as long as they present themselves for work at the correct time and place, then the builder must pay at least basic rates of pay for all the hours concerned. This is known as guaranteed minimum weekly earnings (GMWE). If an operative has no work for a complete week through no fault of their own, then they must receive the basic weekly rate.

Obviously operatives being non-productive in this way can cause the builder considerable financial loss. It is usual when building up the cost of labour to allow for a certain amount of disruption on site due to things like weather, especially when building work is to take place in an exposed area or at times of the year when snow, rain or frost are likely. The allowance is made by a percentage addition to the all-in hourly rate and varies according to the weather pattern for the time of year, the type of work and local conditions regarding things like deliveries.

Loss of guarantee

There is no entitlement to GMWE where the employer is unable to provide continuity of work due to industrial actions, tidal work (WR9) or shift work (WR6).

Proportional reduction

Where an operative is absent for part of the normal working hours due to certified sickness or injury or for one or more days of annual or recognized public holiday, the requirement of the operative to be available for work will be deemed to be met and the payments for GMWE will be proportionally reduced. The proportionate reduction will not apply where the employer authorizes the absence on compassionate or other grounds.

Availability for work

An operative has satisfied the requirements to keep him available for work during normal working hours if they comply with the following conditions:

- He has presented him for work at the starting time and location and has remained available for work during the normal working hours.
- He carries out satisfactorily the work for which he is engaged or suitable alternative work if instructed by the employer.
- He complies with the instructions of the employer as to when, during normal working hours, work is to be carried out, interrupted or resumed.

Temporary lay-off

Where work is temporarily stopped or is not provided by the employer the operative may be temporarily laid off. The operative shall, subject to the provisions of WR 17.4.2, be paid one fifth of the GMWE for the day he is notified of the lay-off and for the first five days of temporary lay-off.

WR 17.4.2 – the payment described above will be paid provided that in the three months prior to any lay-off there has not been a previous period or periods of lay-off in respect of which a guaranteed payment was made for five consecutive days or five days cumulative, excluding the day or days of notification of lay-off. In any such case the operative will not be entitled to a further guaranteed payment until a total of three months has elapsed since the last day of the period covered by the previous payment.

Annual holidays (WR18)

Holidays with Pay shall be as provided for in the Agreement of Holidays with Pay Scheme dated 28th October 1942 and subsequently amended from time to time.

The Holidays with Pay Scheme provides operatives with wages during periods when they are on annual holiday as described in the Working Rules.

The Holidays with Pay Scheme is operated by a management company (Building and Civil Engineering Holidays Scheme Management Ltd, Manor Royal, Crawley, West Sussex, UK) which is non-shareholding, non-profit-making and funded by contributions from all building employers for each of their current operatives. Contributions were made by the weekly purchase of a 'stamp' which was affixed to an employee's 'card' and these stamps were encashed by the employer prior to a holiday period, the money being paid to the employee as 'wages' for the holiday period.

To comply with European Union (EU) employment law, the system has now changed. There is no longer a fixed weekly amount purchasing a 'stamp', indeed there are no longer any stamps. Instead, the employer agrees to pay the management company an amount equivalent to the daily rate of wages for each operative for each day of annual holiday. The money paid is not allocated to any one individual employee and the employer can only claim reimbursement of the amounts paid in. The management company therefore has no control over who receives this money from the employer as payment for holiday 'wages'.

Amounts paid by the employer to the management company must therefore be computed by the employer's wages department and this information passed to the estimator so that they can include provision for them in the calculation of all-in hourly rate.

For example, a craft operative at June 2002 receives an hourly rate of £7.30. Ignoring the Friday of seven hours, that would be a daily rate of £58.40. So for 21 days annual holiday, the employer must set aside a total of £1460. Divided over the productive weeks of the year – 48 – gives a sum of £30.42 which must be put aside by the employer each week for each employee. Table 2.5 gives the amounts for each class of operative. These figures will be used in the examples at the end of the chapter.

The **Winter Holiday** shall be seven working days taken in conjunction with Christmas Day, Boxing Day and New Year's Day, to give a Winter Holiday of two calendar weeks. The dates of each Winter Holiday shall be published by the Council.

The **Easter (Spring) Holiday** shall be the four working days immediately following Easter Monday, to give an Easter Holiday of one calendar week.

It shall be open to employers and operatives to agree that all or some of the 'Winter Holiday' and/or the 'Easter (Spring) Holiday' will be taken on alternative dates.

The **Summer Holiday** shall be two calendar weeks by mutual agreement not necessarily consecutive, to be granted in the 'summer period' – i.e. between 1st April and 30th September. Except by mutual agreement, neither week is to be taken in conjunction with the Easter (Spring) Holiday.

Scottish amendment

The date of the recognized Spring Holiday week will be fixed and published by the Scottish Regional Committee.

By mutual consent and subject to provisions contained within WR18 the Summer Holiday may be taken at a time other than is recognized in the district in which the work is being carried out.

Table 2.5

Name	Job description	Weekly rates based on 39 hours (Current at June 2002)	Per hour	Per day of 8 hours*	HWP provision based on 48 working weeks per annum
General Operative	Carry out general building and civil engineering work	£214.11	549 p	£43.92	£19.22
Skilled Operative	Different levels depending on skill to be undertaken. Work levels are defined in Schedule 1, additional pay per hour is listed.	Skill Rate 4 £230.49	591 p	£47.28	£20.69
		Skill Rate 3 £244.53	627 p	£50.16	£21.95
		Skill Rate 2 £261.30	670 p	£53.60	£23.45
		Skill Rate 1 £271.05	695 p	£55.60	£24.33
Craft Operatives	Carry out craft building and civil engineering work	£284.70	730 p	£58.40	£25.55

* The fact that only 7 hours are worked on Friday can be ignored for purposes of this calculation.

In the event of operatives being required to work during the Spring Holiday, Summer Holiday or Winter Holiday, time and a half shall be paid for normal working hours with exception for time worked on the recognized public holiday day, which will be paid at double time. Overtime work during the holiday period shall be paid at double time.

Public holidays (WR19)

Recouping the cost of wages paid for public holidays involves the calculation of the possible loss incurred as a proportion of wages paid in total. This can be done simply as a percentage of basic hours worked, as follows:

Weeks per year	52
Total annual holidays	4 weeks and 1 day
Working weeks/annum	48 weeks
	×39 hours
	1872
Less 1 day of 8 hours	8
Working year	1864 hours
Public holidays 8 days × 8 hours	64 hours

Thus public holidays represent 3.49 per cent of the working year. This would of course diminish as overtime was worked but only in relation to the productive hours of overtime!

The following are recognized as public holidays for the purpose of this agreement.

Scotland

Christmas Day, Boxing Day, New Year's Day, Spring Monday, the first Monday in May, the Friday preceding the Annual Summer Local Trades Holiday and the Friday and Monday at the Autumn Holiday, as fixed by the competent Local Authority.

Payment in respect of public holidays

Payment for days of public holiday shall be made by the employer to an operative employed by them at the time for each such holiday on the pay day in respect of the pay week in which such holiday occurs, except that payment for Christmas Day, Boxing Day and New Year's Day shall be made on the last day before the Winter Holiday. The amount shall be the equivalent of payment for the normal working hours specified in the Working Hours rule for that day at basic rate.

Payment for work on a public holiday

All hours worked on a day designated as a public holiday shall be paid for at double time.

Scottish amendment

Where the holiday is Autumn Friday, the Autumn Monday shall be paid also for.

Sick pay (WR20)

The Social Security and Housing Benefits Act 1982 gives every qualifying employee the right to receive statutory sick pay when unable to work because of sickness or injury. The Act lays down who will qualify, what minimum amount they will receive and for how long. Basically any operative between 16 and pensionable age qualifies if off sick for four or more consecutive days, which can include Saturdays, Sundays and holidays. The qualifying days for statutory sick pay, and a number of other things such as notification and certification of sickness or injury, have to be agreed by employer and employee. WR20 defines the qualifying days.

Sickness and injury payment

From the estimator's point of view, the most important aspect of WR20 is that it stipulates an additional amount of sickness and injury payment for all qualifying operatives. The qualifications under WR16 for the additional payments may vary from those required by the Act to qualify for statutory sick pay.

At the beginning of 2002 this amount was set at £62.20 per week, there being five qualifying days in the week. Sickness and injury payments are only paid for a maximum of 50 working days; a minimum of four days absence is necessary before any payment is made;

and no payment is made for the first three days of absence. All these rules are further bound in with the accounting period for the holidays with pay scheme.

The bottom line for the estimator is who pays this benefit. The answer is quite simply the employer. The amount paid can vary considerably, but most standard calculations allow five sick days per year. This could be taken as a continuous period of eight days or as five separate periods of four days (3 days without and one day with sick pay – 20 days sick in all). Either way the cost to the employer is the same and would be calculated as follows. From the 52 weeks in the year are deducted 25 days annual holiday and eight days public holiday, resulting in say 45.5 working weeks in the year. Thus the £62.20 for five days sick pay spread over the working weeks is £1.37 per week. With 39 productive hours per week this does not produce a large sum to add to the all-in hourly rate, but it is significant enough not to be ignored. This is the amount we will allow in the examples later in the chapter.

However, this does beg the question of who pays. In the calculation above the client pays. But what if the employer's estimator got it wrong and the employees all had six, seven or more days sick pay? In that instance the employer could lose money. They could of course insure against all sickness and injury payments and build the insurance premiums into preliminaries or overheads or even the all-in hourly rate. They could build in five days as we have shown and insure against any excess; the premiums would be a lot less but the cost to the client at the end of the day might not be significantly less.

Statutory sick pay

Statutory sick pay (SSP) must not be confused with the sickness and injury benefit above. Statutory sick pay is ultimately paid for by central government, although the employer pays the employee in the first instance and then recoups the money from central government.

Statutory sick pay can be payable for up to 28 weeks and there is a system of linking discrete periods of sick leave to give a total of 28 weeks 'continuous' weeks in a period of three years. Among other provisions, linking depends on the periods of sick leave being no more than eight weeks apart. The employer's liability to pay SSP is limited to 28 weeks, after which the employee is transferred onto state benefit.

Statutory sick pay payments are recovered by the employer deducting all amounts paid to employees from the amounts payable for national insurance contributions. If the SSP recoverable is in excess of this amount, the excess may be deducted from the pay-as-you-earn (PAYE) tax paid to the same collector of taxes.

Amounts payable for SSP are set out by the Department of Social Services (DSS) on a weekly basis and vary according to earnings. There are three bands, namely no entitlement, standard rate or lower rate. What is paid can be topped up by the employer to some other agreed minimum, but the employer may only claim back the SSP portion paid out. This top-up effect is achieved by the provisions of WR20.

Relationship of industry sick pay and statutory sick pay

Every employee has the right to receive sick pay under the Social Security and Housing Benefits Act 1982. The Act lays down details of who will qualify, the minimum amount they will receive and for how long.

Scope

This Rule applies to all adult operatives i.e. all operatives aged 18 years and over.

Qualifying days

For the purpose of both this rule and sick pay the 'qualifying days' that shall generally apply in the industry are Monday to Friday in each week.

Amount of payment

An operative who is absent from work due to sickness or injury shall be paid the appropriate proportion of a weekly amount specified by the Council for each qualifying day of incapacity to work. The appropriate proportion for a day shall be the weekly rate divided by the number of qualifying days.

Notification of incapacity to work

An operative must notify the employer on the first day of incapacity to work. Thereafter, the operative will keep the employer informed at regular intervals, not exceeding one week.

Certificate of incapacity to work

The period of absence shall be covered by a certificate(s) of incapacity for work to the satisfaction of the employer. For the first seven days a self-certificate will suffice, additional days must be covered by a certificate(s) given by a registered medical practitioner.

Qualifying conditions for payment

Lists the procedure to be followed by the employee to receive payment. Full details can be found in the Working Rules.

Record of absence

Records of absence must be kept by each employer.

Benefit schemes (WR21)

An operative is entitled to be provided by the employer with cover for retirement benefit, death benefit and accidental injury benefit. Cover for the above benefits are to be provided through payment by the employer of a surcharge on the value of the weekly holiday credits under the Annual Holiday Scheme. There is no entitlement to cover for operatives under the age of 18 or over the age of 65.

Additional voluntary contributions

An operative can opt to add a voluntary contribution in respect of retirement cover. If this is desired the employer shall deduct an amount from the operative's weekly wage and transmit the amount to the Trustees of the Building and Civil Engineering Benefits Scheme in accordance with the rules of that Scheme.

Scaffolders (WR 26)

Scaffolders must be trained in order to carry out their job. Scaffolders employed whole time as such are to be in one of three categories, as defined below:

Trainee Scaffolder – (Rate 4) an operative who can produce one of the following types of the Construction Industry Training Board (CITB) training record card:

- a valid Trainee Scaffolder's card or
- a Basic Scaffolder's card with post-dated validity.

Basic Scaffolder – (Rate 2) an operative who has at least one year's whole-time experience of scaffolding and who can produce one of the following types of CITB training record card:

- a valid Basic Scaffolder's card, or
- an Advanced Scaffolder's card with post-dated validity.

Advanced Scaffolder – (Craft Rate) an operative who has at least two years' whole-time experience of scaffolding and at least one year's experience as a basic Scaffolder and who can produce an Advanced Scaffolder's CITB training record card.

Employer's Contributions for National Insurance

The Department of Social Security publishes a number of leaflets for employers and employees which explain in varying degrees of detail the workings of the National Insurance Scheme. Information may also be obtained from publications such as Croner's Handbook. It should also be noted that the Chancellor of the Exchequer made some far reaching changes in the rules in the budget of 2002. The explanation which follows is not intended to be more than an overview of the rules illustrated in the examples later in the chapter.

First, both employee and employer pay contributions. There is a lower limit for both and and no upper limit for either as a result of the budget changes referred to above. What constitutes a limit is the gross wage earned. This is not necessarily the gross wage used for the calculation of PAYE tax (commonly known as income tax), but the difference is so slight that for practical estimating purposes it will be ignored in the examples. The employer is bound by law to pay both the employee and the employer contributions to the collector of taxes along with PAYE tax, but may deduct employee contributions from gross pay. These contributions for people employed by others are called class I contributions. Special provisions apply to people under 16, pensioners, married women and widows, and employees with more than one job.

When an employer has an occupational pension scheme, the employees are contracted out of the national insurance scheme. For example, plumbers as a whole have negotiated an occupational pension scheme with their employers. The effect is that employees and employers pay lower national insurance contributions. Of course there is the additional expense of the private occupational pension scheme, to which both employers and employees contribute.

National insurance contributions are only payable if the employee earns a certain amount or more, and as previously stated this is the gross wage. If the gross wage exceeds the lower limit it will fall into one of a number of brackets or bands. The lower limits are called earnings limits. Earnings limits, the brackets and the rates of insurance payable can change, normally for a full tax year (6th April of one year to 5th April of the following year).

The rates of insurance payable by both employee and employer are set out in Inland Revenue's DSS booklet 'National Insurance Contributions – Not Contracted-out Tables'.

Table 2.6 *Class I contribution rates by employers for employees not contracted out. (Tax year 6th April 2003 to 5th April 2004)*

Earnings bracket: Weekly gross wage (£)	Employee's rate	Employer's rate
Nil to £75.00	nil	nil
£75.00–£89.00	nil	nil
£89.00–£585.00	11%	12.8%

There are basically two ways in which the amount payable for class I contributions can be calculated. The first is by reference to these tables, which give weekly wages in increments of £1.00 together with the appropriate contribution of employee and employer. The match to wages paid is always to the nearest amount lower than the actual wage; for example, if the actual wage is £140.65, choose rates against £140.00. This method is frequently adopted by wages departments using manual methods of calculating wages.

The second method is to apply a percentage to the actual wage paid. This is the more accurate method and is the one adopted for the examples later in this chapter. It is the method used in computerized payroll systems.

The rates used in the examples later in this chapter are based on the limits, rates and brackets set out in Table 2.6.

Employer's liability insurance

The Employer's Liability (Compulsory Insurance) Act 1969 makes it incumbent upon the employer to insure against injury or disease sustained by an operative while at their place of work.

Premiums for this type of insurance are not expensive and for most trades average out at around £55 per operative per annum. Glaziers, plumbers, painters and decorators and operatives engaged in road works and paving generally attract higher premiums, of the order of £100–£150 per operative per annum. Where woodworking machinery is used, special premiums apply. Woodworking machinery in this context generally does not include hand-held power tools. The majority of policies exclude the use of explosives, tunnelling work or deep excavation, demolitions etc., for which special policies would be necessary. One can insure for literally anything if the premiums can be afforded!

Public liability insurance (against loss or damage caused by employees) is also part of the cost of labour and is generally offered as a package with employer's liability insurance. Although it is not a statutory requirement, many forms of building contract not only require this form of insurance but also stipulate the amount of cover required for any one event. The estimator must therefore ensure that sufficient is built in to the rates to cover these requirements.

Premiums for public liability insurance are based on a rate per operative per annum. They are generally higher than employer's liability insurance and vary in much the same way as those for employer's liability insurance. To give £500 000 cover, the average for the majority of trades would be in the region of £65 per operative per annum. Higher rates can be expected for fencing erectors, floor layers, operatives engaged in roadworks and paving, kitchen installers, landscape gardeners and painters and decorators, somewhere in the region of £80–£120 per operative per annum. To double the cover to £1 000 000 would not necessarily double the premium; usually an additional 20–25 per cent is all that is required. Plumbers and heating engineers are the subject of special agreement; with most companies that means higher premiums.

With both employer's liability and public liability policies there is usually a limit on the number of operatives employed. The average premiums above are for small builders with up to six operatives working on simple domestic building and repair work. Special terms apply when the work involves, for example, deep excavations or scaffolding over a particular height. There are no published rates for these types of work; the employer must obtain quotations from the insurance companies direct.

Because insurance companies generally offer both types of policy as a package, it is convenient to include both in a single percentage addition. At the beginning of 2000 the combined premium averaged 2.5 per cent of the gross wage bill.

Other insurances which the employer may find it prudent to take out – covering such things as insurance of own premises, insurance of the works, and insurance of materials and plant – are not part of the cost of labour but should be included in preliminaries or as an overhead or oncost.

Construction Industry Training Board levy

This is a statutory payment to the CITB to provide training facilities for all operatives and apprentices, although companies with an annual turnover of less than £45 000 are exempt. Payment is now based on a percentage of payroll. Currently this is set at 0.38 per cent of all men 'under contract'.

Redundancy funding

All matters relating to redundancy are covered by the Employment Protection (Consolidation) Act 1978, as amended by the Employment Acts 1980, 1982 and 1988.

In general, payment of redundancy money to an operative depends on the operative being between the ages of 18 and 65 (currently 60 for females) and having been in continuous employment with the employer for at least two years. The nature of employment in the building industry is such that few operatives are in the latter category. Labour tends to move from employer to employer as contracts are awarded and incentive schemes set up to attract operatives. Although many large contracts may run for as long as three years, it would be unusual to have the majority of operatives there for the full contract period.

The scale of payments for redundancy is based upon the weekly wage, which has a maximum of £164.00 where dismissal occurs on or after April 1988, and upon three age bands:

- 41–64 for men and 41–59 for women, for which the payment is 12 weeks' pay for each year of reckonable service.
- 22–40, for which the payment is one week's pay for each year of reckonable service.
- 18–21, for which payment is two weeks' pay for each year of reckonable service.

A year is a full calendar year and reckonable service is limited to the last 20 years of service with the employer. There are rules for the calculation of one week's pay, subject to the maximum mentioned above.

The Acts have always allowed some clawback of redundancy payments by the employer from central government funds. The amount of clawback has steadily diminished with successive amendments and at the beginning of 1990 stood at 35 per cent of the standard redundancy payment but only for employers with 10 or fewer employees. The total of redundancy payments could fall therefore upon the employer, who normally covers for this eventuality by means of a special insurance policy. Approximately 1.5 per cent of gross wages would be a reasonable allowance under current legislation.

Supervision: foremen, chargehands, gangers

The terms 'foreman' and 'chargehand' have been used indiscriminately for so long that it is difficult to know if there was ever any difference in function for these operatives. The common link is of course the supervision of operatives.

The *Oxford English Dictionary* defines 'foreman' as 'principal workman supervising others' and 'chargehand' as 'workman, shop assistant etc. in charge of others'. 'Ganger' is defined as 'foreman of gang of workmen'. These definitions are more or less the same. For the purpose of this section we will consider foreman and chargehand to be synonymous, and to be a principal workman in charge of a squad of craft operatives and labourers. The ganger we will define as the principal labourer in charge of a squad of labourers. There is no terminology used in the Working Rules for trades' chargehands and gangers, respectively.

No matter what we call them, these operatives must receive an additional payment as part of their wages. This additional payment can be paid as a lump sum or an addition to their hourly rate. We will allow in the examples that trades chargehands and gangers will be paid £19.50 per week above the base rate for craft operatives or labourers. This works out at 50 p per hour.

A difference we will impose is that a trades chargehand, while expected to do productive work, may have part of their time taken up with duties on the site which are non-productive. Such duties might include measuring and agreeing bonus for the squad with the builder's surveyor; ordering materials, tools or plant through the site office or through the head office; and checking materials onto site. The ganger will be expected to work productively all the time, the bonus, etc. all being handled by a trades chargehand. This is not entirely satisfactory as it does not cover every situation which might arise. However, it does give a reasonable basis with which to produce sample calculations.

How much time any trades chargehand will spend on supervision will depend on the nature of the job. On large sites with many squads a trades chargehand could spend little time on non-productive work, as there would be a site agent (perhaps with assistants), various clerks ordering materials etc., and full time bonus surveyors and/or clerks. So in this case the chargehand would only agree bonus, take instruction from and report to the site agent, and keep an eye on the squad; in this way, the chargehand could be quite productive.

On small jobs, the chargehand might be the only squad. They would have to cope with everything on the site and so would be left with little time for productive work.

Site agents are of course supervisory, but costs arising from their employment are generally covered as overheads or in the preliminaries. So where does the cost of supervision come from for trades chargehands and gangers? For the trades chargehand it comes from two components of their wage:

- The additional supervisory payment per hour for all basic, overtime and travelling time hours.
- The payment at basic rate of all productive and non-productive hours taken up with supervision.

The amount will be assessed as a percentage of hours; in Example 2.2 the trades chargehand is taken as 75 per cent productive and 25 per cent supervisory.

For the ganger it comes only from the additional payment per hour, as the ganger generally reckoned to be 100 per cent productive.

Bonus, incentive and productivity schemes

These are all devices to promote a higher than normal output from a workforce. For example, bonus is based on the idea that if a piece of work is completed by a group of operatives in say 20 hours rather than 30 hours, then the employer will share the cost of the 10 hours saved with the operatives involved. Therefore, to be able to pay a bonus, the employer must have used realistic outputs when calculating the bill rates. Output is deemed as the number of operative hours taken to complete a unit of work.

So such schemes do not form a direct part of the cost of labour calculation for our purposes.

Excess overtime allowance

This allowance is designed to compensate for the drop in productivity consequent upon fatigue brought about by longer than usual working hours. This occurs especially when these longer hours continue for a number of consecutive days and there is also a prolonged travel period. A drop in productivity could be compensated for by adjusting the output per man in the bill rate. However, this is not a good idea as the estimator might be pricing for a number of squads doing the same work and could confuse travelling squads and local squads. These would have different all-in hourly rates anyway, so why confuse matters by assigning them different outputs? Better to price each squad realistically and leave the outputs alone.

Practical estimating

The explanatory text in this chapter had by necessity to be brief and somewhat general. The reader must appreciate that the subject of labour costs is very wide and each topic can be studied to a greater or lesser depth . Indeed, some of the topics skimmed over here form the basis for complete careers in themselves.

However, we did say right at the beginning that for estimating purposes a few of the details need not be considered and one can generalize for a wide range of others. This, as well as the Working Rules themselves, is best illustrated with a number of examples. We will commence with a very basic situation and proceed to more complex ideas.

Example 2.1

A squad comprising a trades chargehand (T/C), four bricklayers and four labourers is sent to a job 11 km distant. The chargehand will work full time, i.e. supervision will be by the employer. The squad will work the basic 39 hours. Calculate the all-in hourly rate for craftsmen and labourers.

	Craftsman	Labourer
Basic weekly wage = Gross wage in this example	£284.700	£214.110
Supervision:		
T/C 39 hours at £0.50: £19.50 divide by 9 men	£2.167	£2.167
Ganger 39 hours at £0.50: £19.50 divide by 4 men	–	£4.875
	£2.167	£7.042
National health insurance: 11.00% of gross wage	£31.317	£23.552
See Table 2.4 – earnings bracket £89.00 to £585.00		
Holidays with pay	£25.550	£19.220
Public holidays: 3.49% of gross wage	£9.936	£7.472
Sick pay	£1.370	£1.370
Employers liability insurance and redundancy funding: say 4.00% of gross wage	£11.388	£8.564
Wage bill	£368.594	£288.372
CITB levy 0.38%	£1.401	£1.096
Gross cost	£369.995	£289.468
Tea/meal breaks: 4.26% of gross cost	£15.762	£12.331
	£385.757	£301.799
Divide by productive hours (39) worked to get all-in hourly rate	£9.891	£7.738

Supervision

The trades chargehand works or is productive all the time while on site, the supervision being carried out by the employer. Therefore only the plussage paid for being a trades chargehand is chargeable to supervision. This cost is spread over all productive members of the squad, so the divisor is: 1 trades chargehand + 4 craftsmen + 4 labourers = 9 men.

The appearance of a ganger in the calculation may be a surprise. It should be assumed that where a group of labourers is to act independently, one of them at least will get a plussage for being in charge of the others. Note that this is over labourers only, so the divisor is 4, the ganger always being considered fully productive.

In the case of both trades chargehand and ganger, the additional sum they are paid each week should attract National Health Insurance (NHI) contributions from the employer. This would be 11 per cent and amount to just over £1.50 per week; divided by 39 productive hours; it promptly disappears! It is not worth worrying about. There is some rounding up later which more than compensates for this small loss.

Daily fares and travel allowance

The site is within 15 km so no allowance is paid.

National Health Insurance

The gross wage calculated here may not coincide with the definition of gross wage for NHI payments but is near enough for estimating purposes. The amount indicates that employer's contributions are levied at 12.8 per cent. Note that we are not interested in employee contributions.

Other costs

Holidays with pay are the prescribed lump sums and are always included.

CITB levy and sickness and injury pay are included at the amounts calculated.

Employer's liability insurance and redundancy funding are insurance premiums, each based on 2 per cent of gross wage and are lumped together here for ease of calculation as 4 per cent. See Table 2.6.

All-in hourly rate

The total cost per week is divided by the 39 productive hours worked to obtain an all-in hourly rate.

Example 2.2

A squad comprising a trades chargehand (T/C), four craftsmen (bricklayers) and four labourers is sent to a job 34 km distant. The T/C will work 75 per cent of the time, i.e. supervision will take 25 per cent of his time. The squad will work overtime two hours/day Monday to Thursday. Include an allowance of 7.5 per cent for guaranteed minimum weekly wage. Calculate the all-in hourly rate for craftsmen and labourers.

					Craftsman	Labourer
Basic weekly wage					£284.700	£214.110
Overtime:						
	factor	*days*	*hours*	*hourly rate*		
C/man	1.5	4	2	£7.300	£87.600	–
Lab	1.5	4	2	£5.490	–	£65.880
Gross wage					£372.300	£279.990

Supervision: trades chargehand hours

Basic	39	
O/Time	8	
	4	
	51 hours at 0.500	£25.500
Add C/man's gross wage		£372.300
T/C's gross wage		£397.800
NHI: 11.00% of gross wage		£0.438
Holidays with pay		£27.300
Public holidays: 3.49% of gross wage		£13.883
Sick pay		£1.370
Fares & travel allowance for 34 kilometres/daily: 5 at £12.670		£63.350
ELI and redundancy fund: 4.00% of gross wage		£15.912
Wages bill		£520.053
CITB levy at 0.38%		£1.976
T/C's gross cost		£522.029

Tea/meal breaks: 4.26% of gross cost			£22.238
Guaranteed week allowance: 7.50% of gross wage			£29.835
Total T/C cost			£574.102
Less hourly plussage			£25.500
			£548.602
Take 25% as supervisory			£137.151
Add hourly plussage			£25.500
			£162.651
Divide by 8.75 men for craftsmen and labourers		£18.589	£18.589
Ganger: 51 hours at £0.500	£25.500		
NHI: 11.000%	£2.805		
HWP:	£1.750		
ELI & RF: 4.000%	£1.020		
Tea & meal breaks: 4.260%	£1.086		
Guaranteed week allowance: 7.500%	£1.913		
	£34.074		
Divide by 4 men			£8.518
National health insurance: 11.00% of gross wage		£40.953	£30.799
See Table 2.6 – earnings bracket £89.00 to £585.00			
Holidays with pay		£25.550	£19.220
Public holidays: 3.49% of gross wage		£12.993	£9.772
Sick pay		£1.370	£1.370
Daily fares and travel allowance for 34 kilometres/daily			
5 at £12.670	£63.350	–	
5 at £12.580		–	£62.900
Employers liability insurance and redundancy funding: say 4.00% of gross wage		£14.892	£11.200
Wage bill		£549.997	£442.357
CITB levy: 0.38% of wage bill		£2.090	£1.681
Gross cost		£552.087	£444.038
Tea/meal breaks: 4.26% of gross cost		£23.519	£18.916
Guaranteed week allowance: 7.50% of gross wages		£41.407	£33.303
		£617.012	£496.257
Divide by productive hours worked 39 basic hours + 8 O/t equals 47: to get all-in hourly rate		£13.128	£10.559

Hours paid

The total number of hours paid for each man in this example is made up of 39 basic hours, eight overtime hours, four additional hours obtained by taking 0.5 × overtime hours, and five hours of travelling time: this gives 56 hours total. Of these hours, 47 are productive and nine are non-productive.

Daily fares and travelling allowance

The distance of 34 km given in the scenario for this calculation is the distance from the operative's homes to the site as measured under Working Rule 5.2. Five daily allowances are paid in full to each operative.

Supervision

The calculation for supervision for the trades chargehand seems very complex. However, all that is done is the following:

1. We calculate the T/C's gross wage by taking the craftsman's gross and adding the hours worked at the T/C's plussage.
2. We then calculate everything else as would be done for craftsmen and labourers, only using the T/C's hourly rate and the T/C's gross wage.
3. When we have the total cost for the T/C, we separate the additional cost per hour (which is all a supervisory cost) from the remainder (which is only 25 per cent supervisory cost).
4. When we arrive at the latter figure we determine what constitutes 25 per cent and add back the hourly plussage.
5. Finally we divide by the number of men supervised. This is made up of the four craftsmen, the labourers and the 0.75 of the T/C which is productive, giving 8.75.

National Health Insurance

The rate applied is 12.8 per cent (see Table 2.6).

Holidays with pay

Under supervision, both the chargehand and the ganger will have higher hourly wage rates. The chargehand's hourly rate must be fully taken into account when calculating holidays with pay (HWP) provision for him and the plussage paid to the ganger must be taken into account similarly. These work out at:

Chargehand paid £7.80 per hour so his daily rate (8 hours) will be	£62.40
Provision for HWP will therefore be £62.40 × 21/48	£27.30
Ganger's plussage is 50 p per hour, so his daily plussage would be	£4.00
Provision for HWP will therefore be £4.00 × 21/48	£1.75

Guaranteed week allowance

First let us clarify a point. This is not a 7.5 per cent loss of production, costs or anything else. If it was a loss it would not be mathematically correct to add it to our rate; we would expect to add 8.11 per cent.

There is a difference of opinion about the amount to which we should apply the 7.5 per cent. Some schools of thought add it to the basic wage costs only, since that is all that is paid for the guaranteed weekly wage. Others add it to gross wage, as that is the cost to be recouped. It does however, depend on how the percentage has been arrived at in the first place. If the cost of the guaranteed weekly wage is recorded for each trading year or even individual contract, then it will be expressed as a percentage overhead of some fixed sum. The latter can be anything within reason, such as the total basic wage bill, the gross wage bill or the gross wage costs; it makes no difference as long as the estimator knows which.

Example 2.3

A squad comprising a trades chargehand (T/C), three craftsmen (bricklayers) and two labourers is sent to a job 60 km distant. The T/C will work 75 per cent of the time, i.e. supervision will take 25 per cent of his time. One labourer operates a 10/7 concrete mixer, the other a mechanical elevator (power driven hoist). The squad will work overtime three hours/day Monday to Thursday. The squad will travel out to site on Monday, return on Friday evening, and stay in lodgings Monday to Thursday nights inclusive. Include an allowance of 7.5 per cent for guaranteed minimum weekly wage and 5 per cent excess overtime allowance. Calculate the all-in hourly rate for craftsmen and labourers.

					Craftsman	Labourer Skill rate 4
Basic weekly wage					£284.700	£230.490
Overtime:						
	factor	*days*	*hours*	*hourly rate*		
C/man	1.5	4	3	7.300	£131.400	–
Lab	1.5	4	3	5.910	–	£106.380
Labourer's plussage for		57.0 hours at 0.500			–	£28.500
Gross wage					£416.100	£365.370

Supervision: trades chargehand hours

Basic	39	
O/Time	12	
	6	
	57 hours at 0.500	£28.500
Add C/man's gross wage		£416.100
T/C's gross wage		£444.600
Lodging allowance: 4 nights at £18.00		£72.000
NHI: 11.00% of gross wage		£48.906
Holidays with pay		£27.300
Public holidays: 3.49% of gross wage		£15.517
Sick pay		£1.370
Fares & travel allowance for 60 kilometres/daily 1 at £19.310		£19.310
ELI and redundancy fund: 4.00% of gross wage		£17.784
Wages bill		£646.787
CITB levy: 0.38%		£2.458
T/C gross cost		£649.244
Tea/meal breaks: 4.26% of gross cost		£27.658
Guaranteed week allowance: 7.50% of gross wage		£33.345
Excess overtime allowance: 5.00% of gross cost		£32.462

		£742.709
Less hourly plussage		£28.500
		£714.209
Take 25.00% supervisory		£178.552
Add hourly plussage		£28.500
		£207.052
Note: No ganger!		
Divide by: 5.75 men supervised.	£36.009	£36.009
Lodging allowance: 4 nights at £18.00	£72.000	£72.000
See Table 2.6 – earnings bracket £89.00 to £585.00 11.00% of gross wage	£45.771	£40.191
Holidays with pay	£25.550	£19.220
Public holidays: 3.49% of gross wage	£14.522	£12.751
Sick pay	£1.370	£1.370
Daily fares and travel allowance for 60 kilometres/daily		
1 at £19.310	£19.310	
1 at £19.430		£19.430
Employers liability insurance and redundancy funding: say 4.00% of gross wage	£16.644	£14.615
Wage bill	£647.276	£580.956
CITB levy at 3.80%	£24.596	£22.076
Gross cost	£671.872	£603.032
Tea/meal breaks: 4.26% of gross cost	£28.622	£25.689
Guaranteed week allowance: 7.50% of gross wage	£0.312	£0.274
Excess overtime allowance: 5.00% of gross cost	£33.594	£30.152
	£734.400	£659.147
Divide by productive hours worked 39 basic plus 12 O/t equals 51 to get all-in hourly rate	£14.400	£12.924

Rates of wages

The rate of wages applied for the labourers is the Skilled Operative Rate 1 (See Schedule 1 of Working Rules).

Daily fare and travelling allowance

The distance of 60 km given in the scenario for this calculation is the distance from the operatives' homes to the site as measured under Working Rule 5.2. One weekly return fare is reimbursed in full to each operative.

Plussages

The operatives in charge of the concrete mixer and the power driven hoist would both be paid at Skilled rate 4.

No ganger plussage is taken in this example, as the labourers are part of a bricklaying squad and are therefore under the direct control of the trades chargehand.

Holidays with pay

Under supervision, the chargehand will have higher hourly wage rates. The chargehand's hourly rate must be fully taken into account when calculating HWP provision.

Chargehand paid £7.80 per hour so his daily rate (8 hours) will be	£62.40
Provision for HWP will therefore be £62.40 × 21/48	£27.30

3

Mechanical plant

Introduction

Mechanical plant operated by the contractor falls into one of two categories:

- Small plant and tools
- Large mechanical plant and scaffolding.

While either category can be owned by the contractor or hired from a plant hire company, the contractor normally purchases small plant and tools such as hand-held power drills and saws. The contractor may cover the costs of such tools by a percentage addition to the contract. The method for the incorporation of this cost is described in Chapter 4.

When pricing an element in a bill of quantities which includes a plant element the estimator will use either the hire charges quoted by a plant hire company or a rate which has previously been calculated for company-owned plant. The plant rate will normally be quoted as a rate per hour either including or excluding the operator. Simplistically, if it is assumed that a hydraulic excavator with a 1 m^3 bucket can carry out 30 operations per hour and costs £75 per hour, then the excavation costs with this machine will be £2.50 per m^3. There are a number of factors associated with this example, such as how the £75 per hour was derived, the number of operations per hour, the disposal of the excavated material and the balancing of the excavator with the disposal plant. These factors are the subject of this chapter.

Purchase or hire?

Hired plant

Plant can be obtained in a variety of ways, the particular method being a reflection of the contractor's use of the plant. Generally a contractor will own plant which is used daily, whereas plant which is required for a particular operation will be hired on a short-term basis from a plant hire company. This latter category will tend to include specialist plant such as mobile cranes. The hire charge terms of a plant hire company will normally include all costs except fuel, insurance and transporting the plant to and from site. Whether or not an operator is hired with the plant will depend upon the type of plant; obviously for very specialist plant the contractor will hire the plant together with an operator.

Period hire

In situations where a contractor requires plant or vehicles for a longer period but does not wish to own them, there are other arrangements available. For example, a contractor is to commence a two-year project, which requires one van and one truck as site transport. The contractor cannot foresee a use for these vehicles after the end of the contract. One option is to hire as above but an alternative is one of the following.

Contract hire

This is defined as a monthly contract for the hire of the plant or vehicle, which covers everything except the fuel, insurance and operator. Contract hire is provided by specialist contract hire companies who agree to hire for a period that is not necessarily the life of the plant or vehicle.

Leasing

This is similar to contract hire but excludes maintenance of the plant or vehicle. The contractor will pay a monthly charge for the use and will pay for maintenance. The contractor will be able to lease for a period that is less than the life of the plant or vehicle. The Inland Revenue imposes restrictions on the minimum and maximum periods of the lease. Very often under leasing the vehicle is sold at auction at the end of the lease period. In this case there is a reserve price in the lease agreement which is the assumed value that the vehicle will realize at auction. If the vehicle realizes less then the contractor will have to make a cash payment to the leasing company. If the vehicle realizes more then the excess is passed to the contractor.

Contractor's own plant

There are a number of ways in which the contractor can purchase plant.

Outright purchase

This is the most straightforward method. It involves the contractor making a purchase out of cash reserves. The calculation of the real cost of this action is not so straightforward, however, since by investing in a piece of plant the contractor has to forego the opportunity of leaving money on deposit in a bank or even financing another construction project. It can be argued that the cost of outright purchase is the interest, which is lost by not putting the money on deposit in the bank. There is also another dimension: if the money is put on deposit in a bank, then after some years the original sum can still be withdrawn. If however the money is invested in a piece of plant then, as all car owners appreciate, eventually the contractor will possess a worthless heap of scrap. The cost of outright purchase is therefore the interest lost plus the sum of money which would be required to be invested each month to amount to the value of the purchase price over the life of the plant. This latter investment is termed a sinking fund.

Bank or finance company loan

A bank or finance company loan will be made to the contractor for the purpose of purchasing plant. The loan may be for the full or part cost of the plant. The lender may require that the contractor puts in part of the cost from their own reserves and will certainly demand that the loan is secured against the contractor's capital assets as a whole.

Hire purchase

Under a hire purchase agreement the contractor undertakes to repay a loan made by a hire purchase company over a period. The contractor will be required to place a deposit, the amount of which may be governed by legislation. The hire purchase company secures the loan against the plant item being purchased and in the event of a default has no direct recourse to the contractor's assets as a whole.

Lease purchase

Lease purchase is similar to leasing except that the plant is owned by the contractor and appears on the balance sheet. This method of financing takes the total value of the plant down to zero by virtue of the monthly payments and the title to the vehicle is then transferred on payment of a nominal sum. Also the contract may be terminated at any time during the period of the agreement by payment of a termination charge, which is calculated on a sliding scale over the period of the agreement. There is normally in force a legal requirement that the initial payment be a number of months of lease in advance. Under this method the value of the plant is normally written down, i.e. depreciated, over the period of the lease, such that the plant has a nominal value on the balance sheet at the end of the lease.

The plant rate

Mechanical plant for construction work may be hired or purchased in one of the ways described above. Either the plant hire company or the contractor must calculate the cost of plant, which is normally on a rate per hour basis. This section will describe the factors to be taken into account and examples illustrating the procedure are then given.

The plant rate is built up from the following factors:

- Initial cost and finance
- Depreciation
- Interest on capital borrowed
- Life of plant
- Hours worked per annum
- Repairs and renewals
- Insurance and licences
- Fuel, oil and grease
- Inflation.

In addition the estimator must also consider the cost of:

- Transporting the plant to site, which will obviously depend upon the relative positions of the plant and the site
- The operator, which will vary with such factors as travelling time, skill level, bonuses etc. (see Chapter 2).

Initial cost and finance

There are two ways to consider the purchase of plant. The first is that the purchaser has no money and therefore has to borrow to purchase using one of the methods described above. The second is that the purchaser has sufficient money to purchase the plant out of personal reserves.

Borrowing

In the first instance the purchaser has to borrow money from a bank or finance company in order to purchase the particular piece of plant. The cost is therefore the cost of the purchase plus the cost of the loan.

EXAMPLE

Assume that an item of plant has been purchased for £100 000 and that it has a life of seven years. Assume also that the purchaser is able to raise a mortgage, either on that plant for its full value or on other assets such as the purchaser's own office building. The cost to the purchaser would be based upon the mortgage equation, which is

$$R = P[(1 + i)^n i]/[(1 + i)^n - 1]$$

where:
R = the annual repayment;
P = the principal borrowed, in this case £100 000;
i = the interest rate as a decimal;
n = the borrowing period in years, in this case seven years.
If it is assumed that a bank is willing to lend at 13.5 per cent, then the rate of interest as a decimal is 0.135.
Thus the annual repayment is given by

$$R = £100\,000[(1.135)70.135]/[(1.135)7 - 1] = £22\,964$$

Therefore the cost over seven years is £160 748 and the interest paid is £60 748.

Purchasing from reserves

In the second case it is assumed that the purchaser has the money and therefore the cost is a simple £100 000. However, this ignores the fact that the purchaser could invest the money in the bank rather than in a piece of plant. The purchaser therefore suffers the loss of the interest on the investment. Moreover, had the purchaser invested the money in the bank then at the end of the seven years the full £100 000 could have been withdrawn, whereas investing in plant results in the possession of a rusty heap of scrap. In this case, the cost over seven years is the loss of interest plus the contribution to the sinking fund, which accumulates through annual investment to £100 000 over seven years.

EXAMPLE

If it is assumed that the rate of interest for investments is 7.5 per cent, then the annual interest lost is

$$£1000\,000 \times 0.075 = £7500$$

The annual sinking fund equation is

$$R = Ai/[(1 + i)^n - 1]$$

where:
R = the investment expressed as an annual payment;
A = the target accumulated amount, in this case £100 000;
i = the interest rate as a decimal, i.e. 0.075;
n = the investment period, in this case seven years.
Thus the annual contribution to the sinking fund is

$$R = £100\,000 \times 0.075/[(1.075)7 - 1] = £11\,380$$

The total cost per annum is therefore

$$£7500 + £11\,380 = £18\,880$$

Hence the cost over seven years is £132 160.

The rate assumed here is based upon the percentage rate given on deposit in a bank. It may be argued, however, that the percentage rate which should be used is the actual rate of return of investing in the contractor's company, since by buying plant the contractor is losing the opportunity of investing in their own company. The company rate of return for each £1 invested may well be 15 or even 20 per cent and therefore this is the rate which should be used in the above calculation. It will be noted that the taxation implications for the sinking fund are ignored altogether here, whereas in practice they will be taken into account.

It is up to individual preference as to which method is used to calculate cost and finance. However, the former, assuming no capital, has a lot to commend it in terms of its ease of calculation and its logic.

Depreciation

There are two common ways of calculating depreciation for the purposes of the plant hire rate, namely the straight line method and the written down value method. Of the two the straight line method is to be preferred for the purposes of calculating a plant hire rate. It is worth noting here that while all aspects of the plant hire rate may currently be set against corporation tax, the purchase of plant may not. Since the recent abolition of the capital equipment allowance, accountants will set the initial expenditure made on the plant as company expenditure from taxed income. The value of the plant is included in the assets of the company. Depreciation is taken as expenditure in each year that the plant is depreciating.

Straight line method

This method simply involves the estimation of two factors:

- The economic life of the particular item of plant
- The residual or scrap value of the plant at the end of its economic life.

If the plant costs £100 000 and has a scrap value of £10 000 after seven years, then the amount of depreciation per year is calculated as follows:

Cost of plant	£100 000
Value at end of economic life	–£10 000
	£90 000
Years 7	
Depreciation per year	£12 857

Written down value method

Under this method an unchanged percentage rate (commonly 25 per cent) is deducted as depreciation from the previous year's value. This method has the advantage that it more accurately reflects the resale value of the plant at any time, because plant loses the majority of its value in the first year.

Using the same figures as above:	
Capital cost of plant	£100 000
Less 25 per cent depreciation in first year	£25 000
Remaining value of plant	£75 000
Less 25 per cent depreciation in second year	£18 750
Remaining value of plant	£56 250
Less 25 per cent depreciation in third year	£14 063
Remaining value of plant	£42 187
Less 25 per cent depreciation in fourth year	£10 547
Remaining value of plant	£31 640
Less 25 per cent depreciation in fifth year	£7910
Remaining value of plant	£23 730
Less 25 per cent depreciation in sixth year	£5933
Remaining value of plant	£17 797
Less 25 per cent depreciation in seventh year	£4449
Remaining value of plant at end of seventh year	£13 348

This method, although used in accounting, is not generally suitable for deriving a plant hire rate since it presumes a new calculation at the end of each year.

Interest on capital borrowed

The assessment of the total amount of interest has been covered at length above under the heading 'Initial Cost and Finance'. To determine the average yearly value of interest on a loan, the total interest paid is divided by the number of years. In the above example this would be:

£60 748 divided by = £8678.

Life of plant

The anticipated life of plant is generally established by experience. Two factors come into play. The first is the period for which the plant will physically last before the amount of the repair bills become too high compared with the value of the plant. The second is the expected period before the plant is made obsolete by advancing technology. For the purpose of determining a plant hire rate, an estimate on the cautious side is normally the basis for calculation.

Statements from the technical salesmen of plant manufacturers are often a good guide. Such statement might be: 'The plant is guaranteed for one year and all hydraulic seals for two years; most of our operators seem to get about seven years; and there are some machines still operating which are 10 years old'.

This could be taken to mean that there will be little need of repairs in the first two years but following that repairs will increase, until after five to seven years the plant will require to be replaced: it is unusual for it to last 10 years.

Hours worked per annum

Plant is never utilized for every working hour of every working day of the year. Certain types of plant – such as concrete mixers, wheeled tractor-mounted front shovels, backhoe excavators of the JCB3 type, vans and trucks, are likely to be in use for the majority of the year. On the other hand, specialist types of plant such as motorized scrapers and mobile concrete pumps will be restricted in their use by the weather or particular demand.

The number of working hours per annum for a site is approximately 1900. A general purpose excavator may be on hire for 1500 hours per annum, whereas a mobile concrete pump may be on hire for only 800 hours per annum. The exact figures will be drawn from the plant hirer or contractor's own records.

During the on-hire period other factors will reduce the working efficiency of the plant. For example, an excavator may lose the following time, shown as a percentage of available time:

Weather	10
Manoeuvering	10
Breakdowns	5
Operator efficiency	5
General waiting	10
Total	40 per cent

Such factors are taken into account when calculating the effective output of excavating plant per hour.

Repairs and renewals

All mechanical plant suffers from breakdown due to the failure of components. This has two cost effects, the first being the cost of the repair and the second being time lost. The time lost is taken into account in the working output of the plant as described above. Breakdown may be prevented by renewing worn parts during normal periodic maintenance.

The cost of repairs and renewals will vary for different types of plant, but a general figure would be 10 per cent of the purchase price per annum.

Insurance and licences

It is normal for contractors today to have a contractor's all risks (CAR) insurance policy. This will cover, as the name implies, all risks associated with the construction of the building. The premium for this policy is charged as a percentage of turnover (see Chapter 4). The implications for plant are as follows:

1. Employer's liability – this cover is required under an Act of Parliament to cover all employers for their liability in respect of death or injury to workpeople. This cover is extended by CAR policies to cover operatives hired with plant.
2. Public liability – this insurance will cover the contractor against claims by third parties. If, for example, through the negligence of the operative an excavator knocks down a boundary wall and damages a car parked on the other side, then the insurance will pay for claims by the owner of the car.

3. Works insurance – this insurance is primarily for a new building, but can be extended to cover plant for fire, theft and damage. Damage may be accidental or through vandalism. The insurance may also be extended to cover the cost of recovery of incapacitated plant, for example, an excavator bogged down in bad ground.

The above insurances are specific to site-based plant which is not licensed to use the public road. For road vehicles such as vans, trucks and mobile cranes it is necessary to have a vehicle-specific insurance under the Road Traffic Act and also a road fund licence. Since these costs are vehicle specific they are included with the plant hire rate, unlike site-based plant where these costs are excluded.

Fuel, oil and grease

Fuel consumption is generally given in terms of litres per hour. Diesel is available as derv for use on public roads and as gas oil for use on site. The latter is less expensive owing to tax concessions; it contains a red dye and in general it is illegal to use in vehicles taxed for use on public roads. Exceptions are made, e.g. agricultural tractors.

The best source of fuel consumption figures is the site records.

However, a useful rule of thumb is:

- Petrol engine – will consume approximately 0.275 litres of fuel per brake horsepower (bhp) hour
- Diesel engine – will consume approximately 0.2 litres of fuel per bhp hour

From the catalogue of a common hydraulic tracked excavator normally used with backhoe equipment, the maximum power output is given as 130 bhp. When used to excavate a trench, this machine will be working for 5 seconds at full power during a cycle of 20 seconds. For the remaining 15 seconds it is likely to be operating at about half this power. The diesel fuel consumption is therefore calculated as follows:

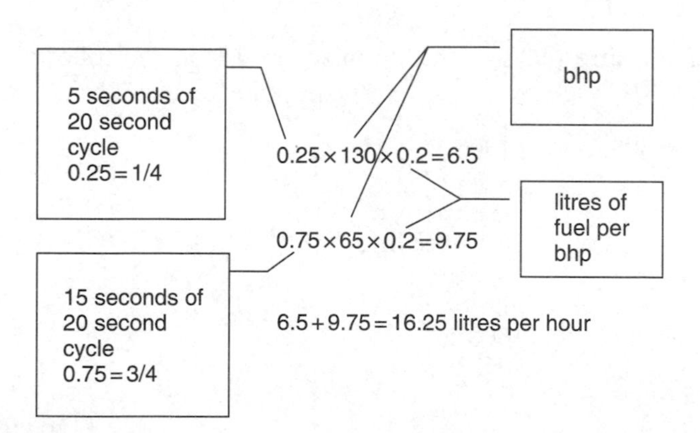

Similar information from the catalogue of a 100 litre (5/3.5) concrete mixer states that the power output is 2.5 bhp. Here the engine will work constantly under approximately the same load and therefore diesel fuel consumption will be $2.5 \times 0.2 = 0.5$ litres per hour. This correlates well with the old approximation of 1 gallon per day.

The cost of lubricating oil and grease can be taken as 10 per cent of the fuel cost.

Tracks and tyres

The cost of maintenance to tracks on plant working on normal ground can be considered to be included with the repairs and renewals figure as calculated above. An exception to this would be where tracked plant is working on rock, for instance in a quarry, where track failure is more common. In this instance repairs and renewals should be 12 per cent of the capital cost per annum.

Tyre life is determined by the tyre size, the loading, the tyre pressure, the type of ground and the skill of the operator. Tyres can represent 15 per cent of the capital cost of a wheeled loader and for this reason various techniques are available for prolonging tyre life. Filling the tyre with flexible foam means that tyre 'pressure' is always correct and production losses due to punctures are removed. Fitting chains to tyres will prolong life, particularly when working on rock. Tyre life for normal and harsh operating conditions is as follows. The figure for harsh conditions should only be used when it is envisaged that the plant will be working continuously in a harsh environment.

Loaders: normal 2500 hours; harsh 1500 hours
Trucks: normal 3000 hours; harsh 2000 hours.

Inflation

A plant rate is calculated at the time that plant is purchased. If no account were to be taken of inflation then during the last year of the life of the plant the rate would be uncompetitively low. Rates are therefore reviewed on an annual basis and increases for inflation are added.

Example 3.1

Calculate the plant rate per hour for a 100 litre (5/3.5) concrete mixer based on the following information:

Data:	Purchase price	£1500
	Interest rate	13.5%
	Engine power	2.5 bhp
	Gas oil	£0.25 per litre
	Life of mixer	10 years
	Scrap value at end of life	nil
	Hours worked per annum	1600
	Working efficiency	90%

Solution: Depreciation: £1500/10 £150.00

Annual repayment on capital borrowed using the mortgage formula:

$$R = P[(1 + i)^n i]/[(1 + i)^n - 1]$$

$$R = £1500[(1 + 0.135)^{10} 0.135]/[(1 + 0.135)^{10} - 1] = £282$$

Total repayment = £282 × 10 = £2820
Interest = £2820 − 1500 = £1320

Interest per annum: £1320/10	£132.00
Repairs and renewals: 10% of £1500	£150.00
Insurance: included as an overhead in the oncost portion of profit and oncost	nil
	£432.00
Divide by hours per annum: £432/1600	£0.27
Fuel: 2.5 bhp × 0.2 = 0.5 litres/hour × £0.16/litre	£0.13
Oil and grease: 10% of fuel	£0.01
	£0.41
Efficiency rating 90%: £0.41/0.9	£0.46

The rate for a 100 litre concrete mixer is £0.46 per hour.

Example 3.2

Calculate the plant rate per hour for a 114 bhp wheeled backhoe excavator based on the following:

Data:

Purchase price	£85 000
Interest rate	13.5%
Engine power	114 bhp
Gas oil	£0.25 per litre
Life of excavator	7 years
Scrap value at end of life	£1000
Hours worked per annum	1400
Working efficiency	60%
Tyre cost	£12 000 per set of 8
Tyre life	2500 hours

Solution:

Depreciation is cost less tyre and scrap values:	
£85 000 – 1000 – 12 000 = £72 000: divide by 7	£10 285.71
Annual repayment on capital borrowed using the mortgage formula:	
$R = £85\,000[(1+0.135)^7 0.135]/[(1+0.135)^7 - 1] = £19\,520$	
Total repayment = £19 520 × 7 = £136 640	
Interest = £136 640 – 85 000 = £51 640	
Interest per annum: £51 640/7	£7 377.00
Repairs and renewals (buckets, teeth, etc.):	
10% of £85 000 – 12 000 (tyres)	£7 300.00
Insurance: none	nil
	£24 962.70
Divide by hours per annum: £24 962.71/1400	£17.83
Tyre depreciation: £12 000/2500	£4.80
Fuel, based on 25% at full power:	
114 bhp × 0.2 × 0.25 = 5.7 litres/hour	
57 bhp × 0.2 × 0.75 = 8.6 litres/hour	
Total 14.3 litres/hour × £0.25/litre	£3.58
Oil and grease: 10% of fuel	£0.36
	£26.57
Efficiency rating 60%: £26.57/0.6	£44.28

The rate for a 114 bhp wheeled backhoe excavator to be used in calculating an excavation rate is £44.28 per hour.

Example 3.3

Calculate the plant rate per hour for an 80 bhp 5 m^3 truck based on the following information:

Data:		
	Purchase price	£35 000
	Interest rate	13.5%
	Engine power	80 bhp
	Derv	£0.50 per litre
	Life of truck	5 years
	Scrap value at end of life	£1000
	Hours worked per annum	1800
	Insurance and road fund licence	£1200 per annum
	Working efficiency	60%
	Tyre cost	£3 000 per set of 6
	Tyre life	3000 hours

Solution:		
	Depreciation is cost less tyre and scrap values:	
	£35 000 – 1000 – 3000 = £31 000: divide by 5	£6 200.00
	Annual repayment on capital borrowed using the mortgage formula:	
	$R = £35\,000[(1+0.135)^5 0.135]/[(1+0.135)^5 - 1] = £10\,073$	
	Total repayment = £10 073 × 5 = £50 365	
	Interest = £50 365 – 35 000 = £15 365	
	Interest per annum: £15 365/5	£3 073.00
	Repairs and renewals: 10% of £35 000 – 3000	£3 200.00
	Insurance and road fund licence	£1 200.00
		£13 673.00
	Divide by hours per annum: £13 673/1800	£7.60
	Tyre depreciation: £3000/3000	£1.00
	Fuel, based on 75% at full power:	
	80 bhp × 0.2 × 0.75 = 12 litres/hour	
	40 bhp × 0.2 × 0.25 = 2 litres/hour	
	Total 14 litres/hour × £0.50/litre	£7.00
	Oil and grease: 10% of fuel	£0.70
		£16.30
	Efficiency rating 60%: £12.91/0.6	£27.17

The rate for an 80 bhp 5 m^3 truck is £21.52 per hour.

Example 3.4

Using the plant described in Examples 3.2 and 3.3 and the information given below, calculate the rate per m^3 for machine excavation of a trench 600 mm wide and the disposal of surplus excavated soil at a tip 2 miles from the site.

Data:

Wheeled backhoe excavator:	
600 mm wide, 400 litre (0.4 m^3) bucket, 1 cycle per 30 seconds	£44.28/hour
Operative	£7.50/hour
Truck 5 m^3:	
average speed to tip 12 mph, tipping time 2 minutes	£27.17/hour
Driver	£7.00/hour
Normal ground: bulking rate	1.1

Solution: Time taken to excavate 5 m^3 of bulked earth:

5 m^3 truck capacity, divided by 0.4 m^3 bucket capacity: 12.5 cycles

13 cycles of excavator to fill one truck, at 30 seconds per cycle: 6.5 minutes to fill one truck

Truck cycle time (minutes):

Loading	6.5	
Travelling	20.0	(4 miles at 12 mph)
Tipping	2.0	
Cycle time	28.5	

Number of trucks required: 28.5/6.5 = 4.38 trucks

A choice is now available between having 5 trucks with the excavator working to capacity, or 4 trucks with the excavator standing idle for a short period.

Each option is now costed.

For 5 trucks the cost per hour is as follows:

Excavator and operative (£44.28 + £7.50)	£51.78
5 trucks and drivers: 5 × (£27.17 + £7.00)	£170.85
Total cost	£222.63

Output per hour (excavator): (60/6.5 min) × 5 m^3 = 46.154 m^3 of bulked earth

Therefore cost is £222.63/46.154 = £4.82 per m^3

For 4 trucks the cost per hour is as follows:

Excavator and operative	£51.78
4 trucks and drivers: 4 × £34.17	£103.68
Total cost	£188.46

Output per hour (trucks): (60/28.5 min) × 5 m^3 × 4 trucks = 42 m^3 of bulked earth

Therefore cost is £188.46/42 = £3.49 per m^3.

Therefore the operation with 4 trucks is more economic.

Thus the rate for excavating a trench 600 mm wide and disposing of excavated material is £4.49 × 1.1 (bulking factor) = £4.49 + 20% overheads and profit = £5.47 per m^3.

Summary

A small contractor whose capital is fully committed may rely solely on hired plant, but as a business grows and workload increases the contractor may decide on outright purchase of specific items of plant for the business. Considerations will be:

- Availability of capital outlay
- State of current and expected workload

- Length of time plant may be required
- Expected rate of utilization
- Proper storage requirements
- Maintenance facilities.

If the contractor can afford the initial and running costs and can predict the utilization factor, then they might decide on plant ownership.

N.B. It may prove advantageous to hire specialized items of plant (excavators, tower cranes, scaffolding etc.) for short periods and own the general types of plant (concrete mixers, dumpers etc.) for permanent usage.

The advantages of plant hire

- Contractor's capital is not locked up in expensive pieces of equipment and cash flow is not affected.
- Hiring costs can be paid monthly as the contractor is paid for work completed in interim payments.
- Contractor does not pay for plant not required nor has any work for.
- Modern and suitable plant is readily available on hire from plant companies at a fixed hourly, daily, weekly or monthly rate.
- Cost of maintenance, repairs and replacement is borne by the plant hire company.
- In some instances the plant hire company will provide the experienced and skilled operative, therefore the contractor does not have permanent operative's wages to pay.
- The contractor has no storage problems when plant is not being used as it is returned to the hire company.

The disadvantages of plant hire

- Effective planning and programming of work schedules is required to minimize expensive idle time on hired plant.
- A suitable machine may not always be available when required, especially at short notice.
- The contractor still has to pay for the hire cost when work is aborted due to bad weather.
- The plant operator does not work for the contractor full-time and so there is no incentive to perform well.
- Plant may not be in good shape to undertake the work, constant breakdown will cause the contractor costly delays.
- Hire rates depend on market forces, which fluctuate and can be a cause of loss to the contractor if the hire cost escalates.

The advantages of buying plant

The contractor loses the following facilities when he or she hires an item of plant in lieu of outright purchasing:

- Ownership of a purchased item of plant is immediately transferred to the acquiring company, which as a company asset can be used as security for finance.
- Outright purchase of an item of plant entitles the acquiring company to capital allowances i.e. tax savings.

- Credit sale – an arrangement whereby title to the item of machinery transfers on purchase but the acquiring company works off the purchase price by installments. This is another good source of tax saving if it can be negotiated.

Cost to own plant

Initial cost:	Interest charges, fixed for short periods
Running costs:	Utilization
	Rate of deterioration
	Cost of plant is high if under-utilized during economic life
	Frequent breakdown increases maintenance costs
	Plant deterioration accelerated if overused with inadequate maintenance
	Obsolescence depends on technological advances

Contractors must ensure plant is economically utilized and does not fall into disuse before the end of its economic life.

The disadvantages of buying plant and hiring plant

The contractor incurs the following additional costs in utilizing plant be it hired or owned:

Cost of transport – From plant depot to site, the extent of this cost depends on the distance from depot to site.
Cost of working base – If tower cranes are employed on site they require suitable concrete base or track to work on.
Cost of erection/dismantling – Contractor incurs the cost of having items of plant (hoists, scaffolding, tower crane etc.) erected and dismantled.
Cost of operators – Machine operators and associated labour such as banksmen, etc. are costs to be considered by the contractor.

4

Preliminaries

SMM7 Section A

Introduction

The preliminaries section of the bill of quantities is the estimator's introduction to the contract. It informs the estimator of the location, size and complexity of the project and gives details of the conditions of contract under which the project is to be completed. The preliminaries section needs care in pricing since it is here that the contractor must or should cover the cost of operating the site under specified conditions and in accordance with the contractor's plan for the progress of the work and for storage and movement of material, plant and site establishment.

> For the purposes of illustration of the pricing of the preliminaries section, all examples will refer to the construction over one year of a 48 m × 56 m factory unit on an existing industrial estate.

The clause numbers following the headings are those of the *Standard Method of Measurement of Building Works*, Section A. A full list in the section is given below, while the areas considered to be of significant relevance have been expanded upon.

Project particulars (A10)

The project particulars give the name, nature and address of the site along with the names and addresses of the client and the various members of the design team.

Tender and contract documents (A11)

This section includes a list of drawings and other documents used in the preparation of the bills of quantities and which will therefore become contract drawings. It is not normal for all the drawings to be sent with the bills of quantities to tendering contractors. Details will be given of the arrangements that the contractor may make for the inspection of drawings.

The site/existing buildings (A12)

Details of the site are given which would include information on site boundaries, existing buildings on, or adjacent to, the site, existing mains/services, underground and over-ground, trees, shrubs, streams, etc.

> *For example*:
>
> The site is situated in Southern Industrial Park, to the west of the existing superstore unit. Access to the site will be by surfaced roads from Albert Parkway and Bishop Road. The contractor will be confined to the area of the site indicated in red on drawing AA/101/89. The contractor is advised to visit the site to ascertain the nature of the site and all local conditions.

Description of the works (A13)

The works are briefly described to give the estimator a feel for the size and scope of the works, the information detailed will cover dimensions and shape relating to each building if not indicated on the drawings, this allows the estimator to carry out rough calculations for scaffolding, etc.

> *For example*:
>
> The works comprise the construction of a factory unit, grid size on plan 48 m × 56 m, with a height to the underside of the roof beam at eaves of 5.72 m. Construction is steel column and beam portal frame, composite panel cladding and aluminium standing seam roofing system. The factory has a two-story office within the main structural envelope. Mechanical and electrical services installation is included in the contract, as are external works comprising roads, car parks, service yards, footpaths and landscaping.

The estimator is now in possession of all necessary basic information. From this point onwards the preliminaries become more specific and can be priced.

The contract/sub-contract (A20)

The form of contract provides a list of contract conditions covering 1, 2, 3, 4 and 5 (below) which gives the estimator the opportunity to price the express requirements of the contract, and to estimate the cost of the risks which are to be borne by the contractor.

1. Schedule of clause headings of standard conditions;
2. Special conditions or amendments to standard conditions;
3. Appendix insertions;
4. Employer's insurance responsibility;
5. Performance guarantee bond/collateral warranties.

This section will assume the use of the *Standard Form of Building Contract, Private With Quantities*, 1998 edition with amendments, issued by the Joint Contracts Tribunal (abbreviated JCT98). Also, for the purposes of this section all clauses will be listed whether or not there is a pricing implication. The JCT98 clause numbers follow each heading.

Clause No.	Title	General comments
1	Interpretation, definitions, etc.	
2	Contractor's obligations.	
3	Contract sum: additions or deductions.	
4	Architect's or contract administrator's instructions.	

5	Contract documents: provision of a master programme.	Although this clause requires the contractor to provide a copy of the master programme, it is probable that this would have been produced by the production planning team within head office and would therefore be costed as a head office overhead. It is thus unlikely that this item would be priced here. Head office overheads are discussed later.
6	Statutory obligations notices, fees and charges.	Where fees are payable arising out of the works described in the bills of quantities then these are to be included within the contract sum. Rates on site hutting may be payable but in this example they are included in the valuation of site hutting.
7	Levels and setting out of the works.	
8	Materials, goods and workmanship to conform to description, testing and inspection.	This is covered more fully under clause A33 (see later).
9	Royalties and patent rights.	If the contractor is required to use a patented process and this is fully described in the bills of quantities, then the contractor should allow here the cost of any licence fee payable.
10	Person in charge.	Traditionally estimators used this item to price the total cost of the site administration. However, bills of quantities prepared in accordance with SMM7 will contain the item 'A40 Management and staff' under which site administration is priced (see later).
11	Access for architect or contract administrator.	
12	Clerk of works.	Facilities for the clerk of works are priced under item A41 (see later).
13	Variations and provisional sums.	
14	Contract sum.	
15	Value added tax.	The conditions relating to VAT are included in a supplement to JCT98 and also in amendment 8. The contract sum and valuations are net of VAT. The contractor can recover VAT on materials purchased and must charge on certain types of building work. At the time of writing only new buildings are zero rated, all other building work being subject to VAT. This could easily change, however. The estimator pricing construction work will always price without the addition of VAT, and all prices in the bills of quantities are similarly without the addition of VAT.

16 Materials and goods unfixed or off-site.

17 Practical completion and defects liability.

Whether or not the contractor declares rates here, the contractor must cover for rectifying defects during the defects liability (or guarantee) period. During this period, which is usually one year, the contractor takes the risk that defects will arise due to materials or workmanship not being in accordance with the specification. In fact most contractors take a broad view of defects and will repair gaps caused by timber shrinkage or plaster cracks even if the timber and the plaster were in accordance with the specification.

An allowance on the factory example might be as follows:

Inspection 4 days at £180	£720
Craftsman 8 × 39 hours at £8.00, say	£2496
Materials, say	£900
	£4116

18 Partial possession by employer.

19 Assignment and subcontractors.

20 Injury to persons and property and indemnity to employer.

21 Injury to persons and property, contractor to insure.

22 Insurance of the works, either contractor or employer to insure.

These insurance clauses are dealt with together here.

The insurance implications of clause 20 can be very complex. The contractor is well advised to discuss with a broker the extent to which the indemnity can be backed by an insurance policy. However, the insurance requirements for the example factory are straightforward and can be summarized as follows:

Employer's liability insurance: All employers of workpeople are required by the Employer's Liability (Compulsory Insurance) Act 1969 to carry insurance to cover claims arising from the death or injury to employees carrying out tasks during their employment. The premium charged for this insurance is based upon the payroll of the company, and therefore the cost can be built into the all-in hourly rate. The insurance is described in more detail in Chapter 2.

Public liability insurance: The contractor is required by clause 21 to carry insurance to cover any claim arising from injury, death or damage to property other than the works caused by the negligence of the contractor's operatives. The premium for this cover is calculated in the same way as for the employer's liability insurance and is therefore incorporated into the all-in hourly rate.

Joint liability insurance: Liability insurance may be required to cover jointly both the employer and the contractor against claims by third parties for damage to property other than the works but which is not caused by negligence. An example of this may be where vibration from piling operations damages a neighbouring structure. The requirement for this insurance will be detailed in the appendix to the contract, and the contractor will usually obtain quotations from a broker dealing with Lloyd's.

Contractor to insure works: Where the contractor is required to insure the works and site materials against all risks, it is necessary for the contractor to first determine the total amount to be insured.

For the factory example this will be as follows:

Contract value	£1 000 000
Allowance for increased costs during construction (see later)	£100 000
Professional fees at 15 per cent	£165 000
Total to be insured	£1 265 000
Premium at say £2.50 per £1000	£316 250

23 Date of possession, completion and postponement.

24 Damages for non-completion.

This is generally a sum per week payable by the contractor to the employer for every week that the building remains incomplete.

25 Extension of time.

The extension of time clause in a contract is important because it determines where loss will fall. If the contract is delayed by actions of the architect or the employer then the contractor is allowed extra time in which to complete, and the damages described above are not payable. The contractor is also able to recover *inter alia* the cost of running the site for the extended period.

If the contractor delays the work by bad management then the contractor cannot claim either the cost of running the site for the extended period or the extension of time which alleviates the payment of damages.

If the contractor is delayed by a matter listed in the contract as being neither party's fault, then the contractor is entitled to an extension of time so that damages are not payable, but is required to cover the contractor's own cost in the tender. This is a risk item, which can be calculated by asking the question: 'What is the percentage probability that this event will delay the contract by one week?' If the answer is 100 per cent then the question has to be asked again with an increased time, i.e.: 'What is the percentage probability that this event will delay the contract by two weeks?'

For the example factory the risk calculation would be as follows:

What is the percentage probability that the following events will delay the contract by one week?

Bad weather	40%	
Strikes	10%	
Non-availability of labour	5%	
Total probability of a one-week delay	55%	
Cost of preliminaries for one week		£7500
Risk cost = 0.55 × £7500		£4125

26 Loss and expense.
27 Determination by employer.
28 Determination by contractor.
29 Works by employer.
30 Certificates and payments.

Determination by employer or contractor (28A).

The conditions contain a provision, which allows the employer to retain a percentage of each payment made to the contractor. This retention fund accumulates during the contract and is paid to the contractor, half at the point where the employer takes possession of the building and half following the satisfactory rectification of defects at the end of the defects liability period.

The effect of this retention is that the contractor is without the use of the retention money for the period of the contract plus the defects liability period.

If, for the factory example, this were to be funded out of bank borrowing, then to the tender sum should be added 3 per cent retention of £1 000 000, i.e. £30 000.

As the retention fund accumulates during the contract from £0 to £30 000, it is assumed that the amount to be funded is £15 000 for one year. For the defects liability period the employer keeps half of the total retention fund, so that £15 000 is to be funded for a further year. The cost to the contractor where bank borrowing attracts interest of 15 per cent is £15 000(1.152 − 1) = £4838.

31	Finance (No. 2) Act 1975: statutory tax deduction scheme.	
35	Nominated subcontractors.	
36	Nominated suppliers.	
37	Fluctuations.	Clauses 38, 39 and 40 give the conditions under which the contractor can recover the increased cost of labour, material, plant, fuel and tax changes during the currency of the contract. The clause, which applies, will be indicated in the appendix to the contract.
38	Contribution, levy and tax fluctuations	Under this clause, often called the fixed price option, the contractor is to include in the tender an allowance to cover the increasing cost of resources. The exception is for those increases resulting from changes in legislation, which are obviously unpredictable. For example, if the contractor gives his operatives a wage increase then this is not recoverable from the employer. If, on the other hand, national insurance contributions increase then the contractor will be able to recover the total cost of the changes from the employer. The contractor is therefore required under this clause to predict the rate of inflation in terms of building costs. The contractor is not liable for the increases in cost to be carried by nominated subcontractors, and therefore the value of this work may be deducted. Domestic subcontracts are often placed during the contract period and therefore an increased cost allowance should be included for these.

In the example factory project the addition is found as follows:

Assume there is 10 per cent inflation over one year, and therefore an average of 5 per cent in the building cost.

Contract value	£1 000 000
Less nominated subcontractors and suppliers	£200 000
	£800 000
Addition to contract 5 per cent	£40 000

39	Labour and materials cost and tax fluctuations.	Under this option the employer elects to take the risk of cost increases, and will reimburse the contractor based upon the proven difference between the quoted price and the invoiced price. Labour rate increases are based upon those incorporated into the Working Rules agreement. The contractor has only to cover non-recoverable increases, which are basically plant and fuel. In the factory example these are insignificant.
40	Use of price adjustment formulas.	Under this option the employer elects to take the risk of cost increases, and will reimburse the contractor based upon the formula method developed by the National Economic Development Office and now maintained by the Property Services Agency. Under this method all increases are covered and the contractor has no need to make an addition.
41	Settlement of disputes: arbitration.	
	Appendix.	The final section under clause A20 is the appendix to the contract. The appendix gives the start and completion dates, the amount of damages payable per week, and so on. One item of significant importance is the base date. The appendix will state the date on which the tender has been prepared for all contractual purposes. This is the date on which the contractor was deemed to be aware of and announced price increases which are to be included in the tender amount. For example, if the base date was set at 4 June 1990 then the contractor is deemed to have included any price increases announced before this date. Therefore if an increase in the wage rate was announced on 3 June its effect is deemed to have been included in the tender. If it was announced on 5 June its effect is deemed not to have been included in the tender. The base date is normally set for ten days before the date for submission of tenders so that the contractor has time for any last minute alterations.

Employer's requirements

Employer's requirements: Tendering/sub-letting/supply (A30)

Employer's requirements: Provision/content and use of documents (A31)

Employer's requirements: Quality standards/control (A33)

Where the design team has specifically requested, the contractor is required to obtain from suppliers samples of materials which will be retained on the site for the purpose of comparison with the materials delivered. Where this involves the contractor in expense, the cost should be added to the tender sum under this item.

> Concrete test cubes may be required to be made from each load of ready mixed concrete delivered to site. For the factory example this would involve the following calculation:
>
> Approximate total volume of *in-situ* concrete: 560 m^3. Delivered by 10 m^3 truck; two cubes from each mix. Number of cubes: $(560/10) \times 2 = 112$ delivered to and tested by local concrete testing station at £35 per cube
>
> Total cost: $112 \times £35 = £3920$

Employer's requirements: security/safety/protection (A34)

Site security is generally described in the bills of quantities by an item similar to the following:

> Adequately safeguard the site, the works, products, materials and plant from damage and theft. Take all reasonable precautions to prevent unauthorized access to the site.

In the factory example this may be undertaken by constructing the fencing described in the bills of quantities at the commencement of the contract and also additional fencing to create a compound for site offices and plant. The provision of the site signboard and the fixing of signs provided by the various consultants are also taken here.

Allowance for repairs to fencing	£1250
Sign board	£350
Compound fencing comprising chain link fence on concrete posts with cranked top and barbed wire:	
100 m at £60 per metre	£6000
Double gates	£600
Alarm system	£800
One-year contract with security company	£12 000
	£21 000

Site safety is a combination of safe working practices, education of operatives and monitoring. These functions are the responsibility of the site management and the safety officer whose costs are taken under clause A40.

The item for protection in the preliminaries section of bills of quantities prepared in accordance with SMM7 covers all work sections and is covered by a general clause:

Employer's requirements or limitations – details stated:	
1. Noise and pollution control	
2. Maintain adjoining buildings	
3. Maintain public and private roads	Fixed Charge
4. Maintain live services	or
5. Security	Time Related Charge
6. Protection of work in all sections	
7. Other	

Unless the contract contains high quality expensive finishes which require specific protection, this clause is a risk item and is not normally priced by contractors.

Employer's Requirements: Specific limitations on method/sequence/timing/use of site (A35)

Employer's Requirements: Facilities/temporary works/services (A36)

Employer's Requirements: Operation/maintenance of the finished building (A37)

Contractor's cost items

Contractor's cost items: management and staff (A40)

The estimator in pricing this item will make use of a form similar to that in Table 4.1.

Table 4.1 *On-site supervision*

Contract period: 52 weeks

Staff	Number staff*	Number cars*	Salary (£/week)	Allowances (£/week)	Total (£/week)
Agent	1	1	600	150	750
Subagent					
General foreman	1		450	200	650
Section foreman					
Trades foreman	1		400	200	600
Engineer	0.1	0.1	750	150	90
Engineer's asst	0.1	0.1	550	150	70
Chainman	0.1		350	125	48
Quantity surveyor	0.2	0.2	800	200	200
QS asst					
Office manager					
Wages clerk					
Timekeeper					
Cost clerk	1		450	200	650
Typist					
Safety officer	0.1	0.1	550	150	70
Plant manager					
Foreman fitter					
Total per week	4.6	1.5			3128
Cost of cars		1.5 × £150			225
Petty cash/expenses	4.6 × £75				345
Stationery	4.6 × £40				184
First aid/laundry	4.6 × £425				115
					3997 × 52
Total cost of supervision					£207844

* Staff/car numbers: 1 day/week = 0.2, 1 week/year = 0.02.

Contractor's cost items: site accommodation (A41)

This item comprises the accommodation required by the contractor in addition to that described in the bills of quantities for the use of the architect or contract administrator and/or the clerk of works. The cost of site accommodation includes (offices, laboratories, cabins, stores, compounds, canteens, sanitary facilities and the like) the two components of weekly hire plus the cost of set-up and takedown. For the factory example the cost is as follows:

Weekly hire	
Staff offices: 1 unit 15 m × 3.6 m at £100/week	£100
Clerk of works office: 1 unit 5 m × 3.6 m at £60/week	£60
Operatives' mess room: 1 unit 15 m × 3.6 m at £100/week	£100
Materials store: 1 hut 9 m × 6 m at £55/week	£55
Portaloo 3 cubicles, 3 urinals: 1 nr at £60/week	£60
Total weekly hire (includes furniture)	£375
Total costs	
Accommodation for 52 weeks at £375	£19 500
Transportation to and from site, and set-up and take-down	£750
Connection of services: 20 hours at £8.00	£60
Total cost	£20 410

Contractor's cost items: services and facilities (A42)

The services and facilities item is to cover the cost of power, lighting, fuels, water, telephone, safety, health and welfare, storage, disposal, cleaning, drying out, protection of work, security, maintain roads etc. These are listed and costed as follows:

Electricity		
Connection charge	£250	
Wiring (included in office unit):		
Electrician 5 hours at £8.00	£40	
Materials	£75	
Cost of electricity 52 at £15	£780	£1145
Water		
Temporary connection	£100	
Water charges at £0.02 per m^3	£400	£500
Telephone		
Connection (site line)	£150	
Connection (clerk of works line)	£150	
Fax/answering machine	£400	
Call/rental charges (site) at £250/qtr 1000		
Ditto clerk of works (provisional sum)	£500	£2200

Local authority rates		
On-site accommodation	£1500	
Fire precautions		
Extinguisher maintenance	£150	
Rubbish disposal		
One skip per week at £60	£3120	
Cleaning		
Offices etc. 52 × 5 days		
2 hours/day at £6.50	£3380	
Site 52 × 5 days, 2 hours/day at £6.50	£3380	
Roads (as necessary) say	£1000	
Professional cleaning on completion	£1500	£9260

Drying out
Dependent upon the time of year, the work of wet trades may need to be dried either by heating or by dehumidification. The example project will not need this.

Small plant and tools
This covers small plant, which the contractor owns as opposed to hired plant. It includes such items as small mixers, 110 volt transformers, power hand tools, picks, shovels, etc. The cost is based upon a percentage of the project cost, say 0.20 per cent of £1 000 000. £2000

General attendance on nominated subcontractors
The bills of quantities measured in accordance with SMM7 will contain this general item for all nominated subcontractors. It allows the contractor to reflect the cost of providing messing and other facilities for the operatives of nominated subcontractors. In the pricing of the example contract this cost has been included in item A41 by an extra allowance in the sizing of facilities.

Site transportation
It is normal to make an allowance for a site van. This will be an all-purpose 1 tonne van for sundry collections and other errands. It will be driven by an employee and may be in the overall care of the general foreman.
1 nr 1 tonne van at £115/week £5798

Contractor's cost items: mechanical plant (A43)

This item allows the contractor to price for large items of plant (including cranes, hoists, personnel transport, earth moving plant, concrete plant, piling plant, paving and surface plant, etc.) the cost of which may not be totally related to one work section. A tower crane, for example, would be used for lifting skips of wet concrete for *in situ* work, steel sections, precast units, etc. The cost of the tower crane would be difficult to distribute amongst these items and therefore its cost is better reflected here.

Also, in situations where work is marked as provisional, the contractor may wish to price the plant associated with that work under this section. For example, the bills of quantities may indicate piling comprising 100 driven precast concrete shell piles provisionally 15 m long. If the project demanded two piling rigs for three weeks at a cost of £3500 per week each, then the cost may be included as a lump sum of £21 000 in this section or alternatively as £14 per metre of piling.

The problem with the latter alternative is that the period that the rigs are required on site is related as much to the number of piles to be driven as to their total length. If each pile were driven 14 m instead of the 15 m anticipated, this would be unlikely to have any effect upon the total time that the rigs were required on site. However, had the pricing been based upon £14 per metre of piling the contractor would have received £1400 less income.

Contractor's cost items: temporary works (A44)

This item is provided in the bills of quantities for the pricing of any temporary works which is not to be included as an element of the bill rate. Temporary works which may be included are temporary roads, temporary walkways, access scaffolding, support scaffolding and propping, hoardings, fans, fencing, hard standing and traffic regulations. In the example factory, scaffolding is to be considered as an item under this heading. Scaffolding is most commonly provided by a specialist scaffolding company, which will quote for the erection and dismantling of scaffolding plus a hire rate per week.

It is assumed for the purposes of the example that scaffolding will be provided by the contractor for the brick office structure to be constructed within the framed building. From the information given in the description of the works it is assumed that a freestanding scaffold will be required 48 m long × 4.5 m high × 1.2 m wide.

The names given to the various scaffolding components are as follows:

Standards Vertical tubes 1.2 m apart, at 2 m centres along the face and back of the scaffold.
Ledgers Horizontal tubes running the length of each side of the scaffold, at 1.5 m vertical centres.
Putlogs Horizontal tubes across the width of the scaffold, at 2 m centres horizontally and 1.5 m centres vertically.
Braces Diagonal tubes across the face and back of the scaffold.
Boards Decking and toe boards.

The costs are calculated as follows:	
Quantities	
Standards: [(48/2) + 1] × 2 = 50 × 4.5 m high	225 m
Ledgers: (4.5/1.5) × 2 × 48 m long	288 m
Putlogs: [(48/2) + 1] × 4.5/1.5) × 1.5 m long	113 m
Braces: say	72 m
Total tube	698 m
Fittings: say 4 per putlog (300) + 60 = 360	
Boards (225 mm wide): 5 boards + toe board = 6 × 48 m = 288 m	
Erect and dismantle time	
Tube: 698 m at 0.1 hours/metre	70 hours
Fittings: 360 at 0.15 hours each	54 hours
Boards: 288 m at 0.06 hours/metre	17 hours
	141 hours

Cost: erection and dismantle plus hire six weeks	
Labour: 141 hours at £8.00 per hour	£1128.10
Hire: Tube 698 m at £0.05/m per week × 6 weeks	£209.40
Fittings 360 at £0.02 each per week × 6 weeks	£43.20
Boards 288 m at £0.08/m per week × 6 weeks	£138.24
Transport: say 5.5 tonnes at £30 inc. loading and unloading	£275.00
Cost of scaffolding	£1793.84

Work/products by/on behalf of the employer (A50)

Nominated subcontractors (A51)

SMM7 requires that all nominated subcontract work be described in the preliminaries section. The estimator is required to price two items in connection with this work, namely profit and special attendance. An example is as follows:

Prime cost Allow the prime cost sum of £14 500 for steelwork to be fabricated, supplied and erected by a nominated subcontractor	£145 000
Profit Add for profit 5 per cent	£7250
This item allows the estimator to add for profit at a different rate to the profit on builder's work. If the profit is priced as a percentage it will be adjusted proportionally to the adjustment of the prime cost sum.	
Special attendance Add for special attendance: lorry mounted 7 tonne capacity telescopic crane and operator for 10 days during the construction of the steel frame. This facility is to be provided by the contractor for use by the nominated subcontractor. The cost which the estimator will include here is 12-day hire at £300 per day	£3600
Total for A51	£155 850

Nominated Suppliers (A52)

Work by Statutory Authorities/Undertakers (A53)

Provisional work items (A54)

SMM7 sets out rules for provisional sums, which are either:

- Defined, in which case the estimator should consider whether there is a preliminary cost implication in the definition and include any cost here; or
- Undefined, in which case no additional cost is included in the preliminary bill. All preliminary costs associated with the expenditure of an undefined provisional sum are to be included with the valuation of work when the work is carried out.

Dayworks (A55)

Under this item the estimator is required to enter the percentage additions to the allowances made for labour, plant and material. The percentage additions are to cover the contractor's costs over and above those defined in the *Definition of Prime Cost of Daywork* carried out under a Building Contract, published by the RICS and the BEC.

It should be noted that under the conditions of JCT98, daywork may only be used to value work, instructed by the architect or contract administrator as a variation, where the work may not be measured.

Collection of example costs

The value of the preliminaries for the example factory is summarized in Table 4.2.

Table 4.2 *Preliminaries for factory example*

Practical completion and defects liability		£4116.00
Insurance of the works		£3162.00
Extension of time allowance		£4125.00
3% retention financing charges		£4838.00
Fluctuations allowance		£40 000.00
Concrete test cubes		£3920.00
Site security		£21 000.00
Management and staff		£207 844.00
Site accommodation		£20 410.00
Services and facilities:		
Electricity	£1145.00	
Water	£500.00	
Telephone	£2200.00	
Local authority rates	£1500.00	
Fire precautions	£150.00	
Rubbish disposal	£3120.00	
Cleaning	£9260.00	
Small plant and tools	£2000.00	
Site transportation	£5798.00	£25 673.00
Cost of scaffolding		£1793.84
Nominated subcontractors		£155 850.00
Total cost of preliminaries		£492 732.34

5

Excavation and filling

SMM7 Section D2

Site preparation

This section is concerned with the removal of trees, tree stumps, bushes, scrub, undergrowth and hedges. A tree or tree stump is defined in SMM7 as having a trunk of at least 600 mm girth at 1 m above ground level or at the top of the stump. The assumption can only be made that anything smaller is a bush.

Guidelines for estimating for the removal of trees and stumps might include the following:

- Trees: Time for squad to fell; hire of chain saws; fuel and oil; sharpening; axes, wedges, pince bars and mauls; protective clothing; time to cut up and burn the slash; disposal of the main limbs and trunks to a tip or to a firewood merchant.
- Stumps: Time to excavate around larger stumps by hand, or by machine if numbers or size warrant; men excavating or hire of machine and operative; disposal of stumps to tip (firewood merchants seldom want to know about stumps).

Bushes, scrub and undergrowth can be removed by hand using chain saws or scrub saws and burning the residue. Machines can also be used to 'bulldoze' the scrub off the ground into piles for burning. Once again the choice depends on the size of the site to be cleared and the density of the growth. Burning may not be allowed in smokeless zones or because of fire hazards nearby and disposal to a tip may be the only solution.

Example 5.1

Felling a tree, girth at 1 m above ground level say 2 m, burn slash, and remove trunk and main limbs off site.	Nr

The only way to price the item is first of all to look at the quantity involved, and from this and the item description, to then assess the likely squad size and what is required by way of plant. An isolated tree or group of up to, say, five trees can be dealt with as follows:

Felling 5 trees, cutting up, burning, loading onto lorry:		
3 labourers for 2 days of 8 hours at	£6.50	£312.00
Hire of chain saw, nr 2, for 2 days at	£8.00	£32.00
Lorry with MOL and driver for 2 days at	£45.00	£90.00
* Charges at tip: 3 loads at	£12.00	£36.00
		£470.00

* This charge does not include Landfill Tax – see notes on this tax in the section on Disposal.

Profit and oncost 20%	£94.00
	£564.00
Rate per tree	£112.80

A lorry with mechanical off-load (MOL) has a hydraulic crane mounted on the body. The crane can be fitted with a variety of devices for lifting different kinds of loads, e.g. pallets, round logs, slings for beams, and hooks.

Example 5.2

> Remove stump and root of tree, girth 2 m at 1 m above ground level, including refilling hole with soil and disposal of stump off the site. Nr 5

There are two ways in which stumps can be removed: the first is to dig them out and the second is to pull them out. The latter is cheaper but really requires the stump to be left with a good length of trunk intact. In this instance a powerful machine, probably tracked, is attached to the top of the stump with heavy chain. The machine literally pulls the stump out, exerting considerable leverage by virtue of the extended piece of trunk. Of course one cannot assume that, just because the tree stumps are left with a length of trunk, they can be pulled out economically. There has to be a tracked excavator on the site doing other things! The rate for stump removal would be very high if a bulldozer was to be brought onto site to take out five tree stumps such as we have described.

Excavating stumps is most easily accomplished with a backhoe on a machine such as a JCB. Having weakened the hold in the soil, the JCB might be able to pull the stump out, or the main roots may have to be cut before the machine can dislodge the stump. Excavation is treated as follows (assuming that there is an excavator on site):

Root spread approximately 1.50 m radius
Root depth approximately 1.25 m
Volume of excavation approximately 8.84 m^3

Excavation for 5 stumps: 44.2 m^3 at 10 m^3/hour at £96.00	£424.32
Backfill excavation: 44.2 m^3 at 15 m^3/hour at £96.00	£282.88
Loading onto lorry and disposal: lorry time included in Example 5.1	–
Charges at tip: say 5 loads at £12.00	£60.00
	£767.20
Cost per stump	£153.44
Profit and oncost 20%	£30.69
Rate per stump	£184.13

Lifting turf for preservation

Natural sites seldom have turf worth preserving, unless they happen to be one of the rare pieces of ground with some special soil where the turf can have an almost lawn-like quality.

Turf for preservation is most likely to be encountered where the ground has been previously cultivated and is still in reasonable condition. If the area is large, it might be worth considering obtaining quotations from a specialist landscape gardener to bring in a machine to lift and roll the turf. Lifting turf by hand is a skilled operation using a special spade and should not be undertaken by just anyone. The thickness of the turf is critical, as is the rolling, or the stacking of flat turves. If the estimator is in any doubt as to the feasibility of his own men coping with the work, then again a reliable landscape gardener should be asked to quote for the work.

Excavation: general remarks

The rules of measurement for SMM7 Section D2 have been drafted with machine excavation in mind as the principal method. Any excavation which the estimator judges can be carried out only or better by hand must be priced within the constraints of these rules. The implications of this for both the estimator and the person preparing the bill of quantities are discussed later in the chapter.

Excavation is measured in eight classes or types of excavation:

1. Top soil for preservation;
2. To reduce levels;
3. Basements and the like;
4. Pits (give total number);
5. Trenches up to 300 mm wide;
6. Trenches over 300 mm wide;
7. For pile caps and ground beams between caps;
8. Benching sloping ground to receive filling.

For the first class, state the average depth. For the remaining classes, give the maximum depth in stages: up to 250 mm deep; up to 1 m deep; up to 2 m deep; and so on in stages of 2 m. In addition, a starting level for the excavation should be provided where the particular class of excavation commences beyond a depth of 250 mm below existing ground level.

The subclassification into depth stages with a given starting level is a blanket requirement of SMM7.

The class 'To reduce levels' has been retained from SMM6 and is assumed to be the same, i.e. the excavation takes place *above* the starting level. The blanket rule that a starting level is stated when the excavation starts beyond 250 mm below the ground level obviates any unnecessary provision of a starting level for reduce level.

The last class, 'benching sloping ground to receive filling', relates to the practice of excavating a permanent unsupported side slope down which there are horizontal steps which, when filled over with imported material, prevent that material slumping to the bottom of the excavation. Figure 5.1 illustrates the principle. Again the necessity to state starting levels and give depth stages can be questioned. If the excavation is a 'hole in the ground' with permanent side slopes stepped and filled over, then there is a starting level and a depth to consider. However, the most common occurrence of this type of work is in large cuttings such as those encountered in motorway construction, where the excavation is entirely above the starting level and this last is the original ground level!

The previous two paragraphs have been critical of SMM7. Whether these criticisms are justified or not, they serve to illustrate that a quantity surveyor will interpret these rules and present the bill for pricing in the firm belief that his item descriptions and all the other

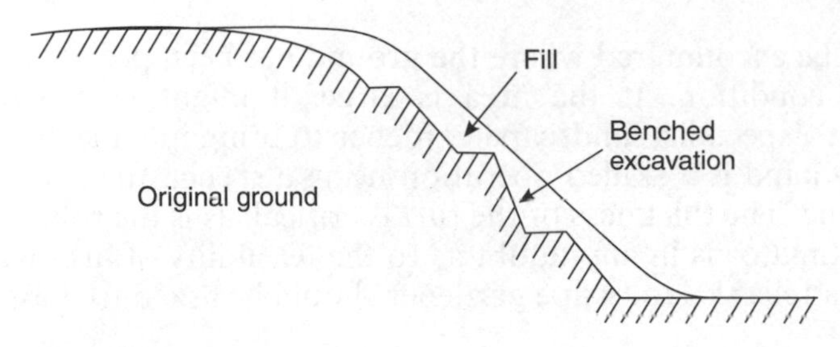

Figure 5.1 *Benching sloping ground to receive filling*

supporting documentation really do represent the work properly. The estimator must be certain that he is pricing the intended work.

There is no requirement in SMM7 to state in the bills of quantities whether or not the work is to be done by machine or by hand. However, the Measurement Code published with SMM7 does state that 'Information provided in accompanying drawings or in descriptions should identify those circumstances where it may be difficult or impractical to carry out excavation by mechanical means'. The final judgement must be made solely by the estimator. To enable him to make it, the bill should clearly state whether the work is inside an existing building, under water (and whether this is river, spring or tidal) or in compressed air (SMM7, General Rule 7). A study of the drawings and the preambles will elicit further information, as will a visit to the site. However, there are further considerations which are known to the estimator and depend on his experience and the technical capabilities of his company. These are all discussed below.

Hand excavation

Not all excavation is possible using machinery. Nor is there one type of machine which is suitable for all classes of machine excavation.

For example, excavation of foundation trenches using a backhoe on a wheeled tractor unit will not give accurate alignment of the trench, and the teeth on the bucket will disturb the bearing surface of the trench. Further, it is not possible to dig to a precise level and depth with sufficient accuracy. The backhoe will therefore take out the bulk of the excavated material – say 90 per cent and the trench will be trimmed to level for the foundation by hand. Hand trimming to width may also be required where there is extensive timbering.

Hand excavation may also be necessary for small isolated pits for pads and manholes, excavations for foundations to entrance steps, and so on, either because it is uneconomical to bring a machine to the work or because access is restricted. In the latter instance one could also consider work inside an existing building or in a back garden or enclosed yard or area.

On the subject of trimming excavations by hand, it should be noted that no matter how the excavation is done, the smaller and more confined types of excavation have a higher proportion of trimming to volume of excavated material. Generally in the past this has been taken into account by using lower outputs per hour when calculating the overall rate, which is based purely on volume. However, SMM7 and previous editions have included provision for the preparation of excavations to receive concrete foundations.

Depth and throw

Excavating by hand – even only the trimming of machine-excavated trenches, basements or whatever – brings with it the problem of reconciling the depth stages in SMM7, which are set to suit machines, with the lift and throw achievable by a man with a shovel. The norm for hole depth for man and shovel is 5 ft, which is translated loosely into 1.5 m. This is generally held to be shoulder height. Figure 5.2 and the following paragraphs explain the reduction in outputs per hour with increasing depth.

Refer first to situation A in Figure 5.2. The man in position 1 has dug out the earth and has piled it at the side of the excavation. As he has gone down, the pile of spoil at the side has risen, thus increasing further the required throw. It is for this latter reason as much as depth that a limit of 5 ft was set for depth.

Then refer to situation B. The man in position 1 continues to dig down but can no longer throw the spoil clear of the excavation and maintain a reasonable output. Reasonable output is the important point at this transition between depth stages. Of course he could throw clear if the excavation were to stop at say 1.65 m, but the effort required to clear the spoil from the excavation over the growing pile at the surface would slow him down, and that would be unacceptable were the digging to continue for even greater depths. So to keep output up, the man erects a platform or staging 1.5 m down (position 2) and puts the spoil on the platform as he continues to dig down to 3 m. Either he or a second man works off the platform shovelling the now loosened spoil up on to the side of the excavation. We have already stated that the pile of spoil at the side of the trench was making it difficult for the man excavating up to 1.5 m down, so it must also be increasingly difficult for the man shovelling from the platform. What we require to have is the loose spoil cleared back from the side of the excavation (position 3). Again this can be the one man or it could be a second man alternating between staging and starting level.

Finally, situation C shows the saga continuing down to 4.5 m with a second staging, and the need to man both stagings and the surface to clear away.

The man at position 3 can of course do a number of operations. He can merely shovel the spoil back to make room for more and make ready for backfilling in the case of pipe

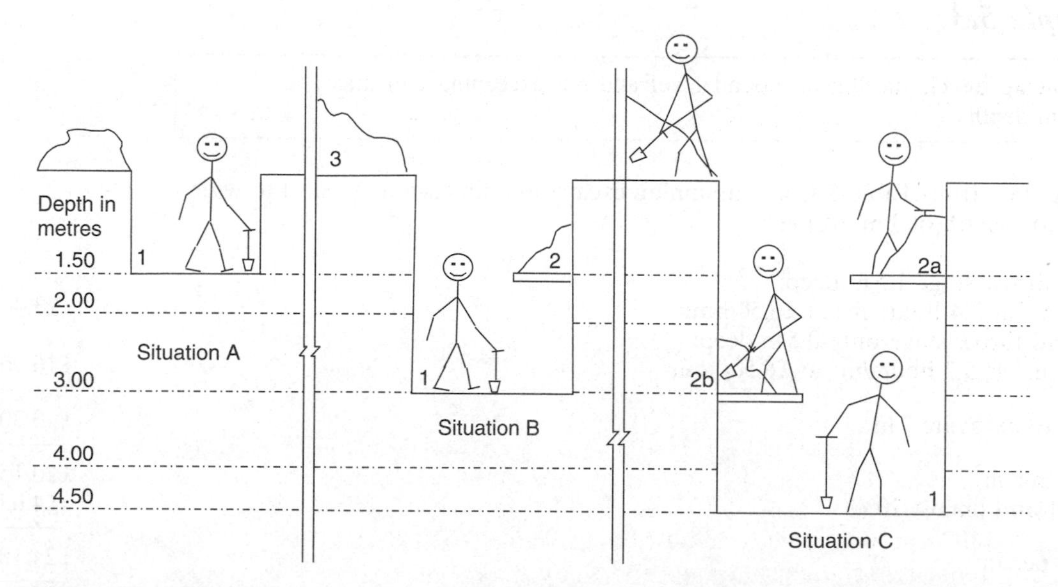

Figure 5.2 *'Throw' in minimal excavations: Table 5.1 gives the times taken to get out 1 m³ spoil*

Table 5.1 *Hours to excavate 1 m^3 by hand*

Depth stage (m)	To excavate and get out	To throw one stage	To clear sides	Total
>1.5	2.4	0.0	0.0	2.4
1.5 to 3.0	2.4	1.4	1.4	5.2
3.0 to 4.5	2.4	2 × 1.4	1.4	6.6
4.5 to 6.0	2.4	3 × 1.4	1.4	8.0

trenches. He might also segregate the spoil according to the different strata being encountered, some being suitable for backfill round pipes; vegetable soil might be kept separate, and the remainder used as general backfill after a certain level. He might load everything into wheelbarrows, dumper trucks, skips or hutches.

Assuming the man at position 3 is clearing back ready for backfilling, then typical outputs are as shown in Figure 5.2. The man getting material out from the bottom takes almost twice as long to handle one cubic metre as those shovelling off the stagings or clearing back at the surface. No matter how many men are required, what the squad size is or what the disposition of the workers can be, the times taken to get out one cubic metre of spoil can be calculated as shown in Table 5.1. The outputs used in Table 5.1 are averaged from Tables 5.3 and 5.4, and assume ordinary soil conditions as defined in Table 5.2.

The hard part about all of this is to reconcile these ergonomically selected depth stages with those given for machine excavation in SMM7. Figure 5.2 has these depths shown as dashed lines across the illustration.

There follow two examples of rates for hand excavation, using the outputs calculated in Table 5.1.

Example 5.3

Excavate trench starting at ground level and not exceeding 2 m maximum depth.	m^3

Using data from Table 5.1, and assuming excavation dimensions to be 1 m wide, then for length of 1 m on plan:

First throw stage 1.5 m deep: 1.5 m^3 at 2.4 hours/m^3 at £6.50/hour	£23.40
Second throw stage only 0.5 m deep: 0.5 m^3 at 5.2 hours/m^3 at £6.50/hour	£16.90
Cost to excavate 2 m^3	£40.30
Cost per m^3	£20.15
Profit and oncost 20%	£4.03
Rate per m^3	£24.18

Example 5.4

Excavate trench starting at ground level and not exceeding 4 m maximum depth.	m^3

Using data from Table 5.1, and assuming excavation dimensions to be 1 m by 1 m on plan:

First throw stage 1.5 m deep:	
1.5 m^3 at 2.4 hours/m^3 at £6.50 / hour	£23.40
Second throw stage 1.5 m deep:	
1.5 m^3 at 5.2 hours/m^3 at £6.50 / hour	£50.70
Third throw stage only 1 m deep:	
1 m^3 at 6.6 hours/m^3 at £6.50 / hour	£42.90
Cost to excavate 4 m^3	£117.00
Cost per m^3	£29.25
Profit and oncost 20%	£5.85
Rate per m^3	£35.10

In both these examples, the staging necessary could be priced in with these rates or included with the items for earthworks support.

Soil conditions

To complicate matters further, the estimator will also be required to consider that the material being excavated may vary in consistency from liquid to solid, that is from silt or sand below water level to hard rock such as whin or granite.

Obviously some materials are easier than others to remove, and the greatest differences are to be found in hand excavation. Table 5.2 gives a range of multipliers which may be applied to an output for normal soil to obtain outputs for various types of hand excavation in a range of soil types.

Soil conditions should be defined in the preambles section of the bill for excavation and filling, but the site visit should be used as check on all of these points. Indeed, if no trial pit or bore information is given the estimator might consider it prudent to obtain permission to have this done.

Outputs

Table 5.3 gives a range of outputs for various classes of hand excavation in ordinary soil. The variations in the figures are a consequence of a number of factors, the more important having been identified in Table 5.2.

Table 5.4 shows sundry outputs for various hand excavation tasks.

Examples 5.5 to 5.10 are based on an imaginary building. A section of the earthworks is shown in Figure 5.3. The difficulty multipliers are obtained from Table 5.2 and the man hours per m^3 from Tables 5.3 and 5.4.

In these examples the volume of material to be excavated is calculated for an area of unit length and unit width, i.e. 1 m × 1 m. The depth is, in the case of the reduce level digging, the

Table 5.2 *Multipliers for difficulty due to nature of ground and type of hand excavation*

Soil type	Multiplier (× hours/m^3)	Excavation type	Comment
Vegetable soil	1.11–1.33		Because of small volume for a given area
Sand	1.25–1.52	Reduced level, basement, trench, pit	Depending on degree of compaction and moisture content, a shovel will not hold as much sand as soil
Ordinary	1.00	Reduced level, basement, trench	Any soil capable of being removed with a shovel and minimal use of the pick
	1.11–1.25	Narrow trench, pit	Higher proportion of trimming
	1.11–1.33		
Heavy clay	1.25–2.00	Reduced level, basement, trench, pit	Graft or pneumatic spade; trimming is almost assured with accurate use of tools
Soft rock	1.43–3.00	Reduced level, basement, trench, pit	Heavy use of pick or wedges; trimming is almost assured with accurate use of tools
Hard rock	3.00–5.00	Reduced level, basement	Wedges or compressor and pneumatic drill
	5.00–10.00	Trench, pit	Trimming difficulty and confined space

Table 5.3 *Range of outputs for various classes of hand excavation in ordinary soil (hours/m^3)*

Oversite <250 m	1.25–1.50
Basement and the like	2.00–2.25
Trenches (depending on width)	2.00–2.75
Pits (depending on size)	3.00–4.50
Reduce level	1.75–2.00
Benchings	1.75–2.00

average depth. For example, in Figure 5.3 the shallowest portion of reduce level digging starts at zero and rises to 0.25 m deep, giving an average depth of 0.125 m. The volume excavated is therefore $1\,m \times 1\,m \times 0.125\,m = 0.125\,m^3$. Similarly, the deepest portion of reduce level digging starts at 1 m deep and rises to 1.2 m deep, giving a total volume of $1\,m \times 1\,m \times 1.1\,m = 1.1\,m^3$ (Example 5.7).

Table 5.4 *Sundry outputs associated with hand excavation*

Volume outputs (hours m^3)	
Wheel 50 m and deposit	0.60–0.80
Additional 50 m	0.60–0.80
Spread and level	0.40–0.50
Re-excavate from spoil heaps	1.25–2.00
Return fill and ram	1.25–1.75
Lift one stage	1.25–1.50
Clear from side of excavation	1.25–1.50
Clear from side of excavation and segregate from backfilling	1.40–1.80
Barrowing and filling hardcore in bulk	0.80–1.00
Barrowing and filling hardcore in beds <250 mm	1.15–1.30
Mechanical ramming bulk fill in 300 mm layers	0.40–0.60
Fix and withdraw timbering	7.00–10.00
Area outputs (hours/m^2)	
Grading/levelling trench bottoms	0.15–0.20
Mechanical ramming hardcore in beds	0.12–0.13
Spread and consolidate blinding	0.05–0.10

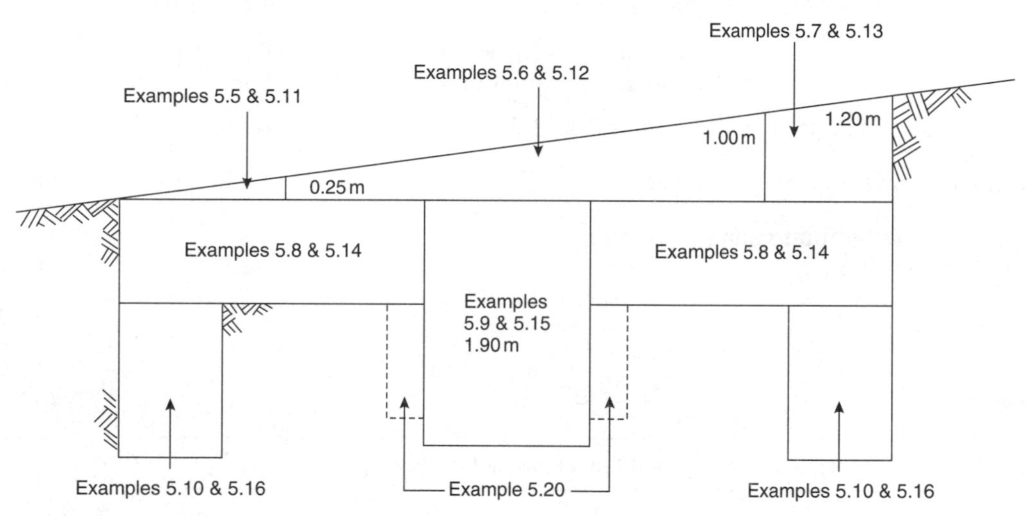

Figure 5.3 *Cross-section through a typical excavation*

Example 5.5

Excavate to reduce level, starting at ground level and not exceeding 0.25 m maximum depth.	m^3

SMM7 has already been criticized for requiring depth stages and starting levels in items for reduce level excavation. However, the estimator knows that this work is all above the starting level and that the subsoil is a compacted sand/gravel. In this instance, the get-out part of the excavation will require barrow runs, mechanical barrows or dump trucks. Provision of barrow runs will be included in the general items in preliminaries, and the additional labour is reflected in the choice of difficulty multiplier.

Assume that we will excavate unit length and unit width at any time. Average depth (0 + 0.25)/2.

	Volume excavated	Difficulty multiplier	Man hours	Rate per hour	Cost
Max. throw from 1.5 m deep	0.125	1.11	1.5	£6.50	£1.35
Total volume excavated	0.125			Total cost	£1.35
Dividing total cost by total volume = cost/m^3					£10.82
Profit and oncost 20%					£2.16
Rate per m^3					£12.99

Example 5.6

Excavate to reduce level, starting at ground level and not exceeding 1 m maximum depth.	m^3

Assume that we will excavate unit length and unit width at any time. Average depth (0.25 + 1.00)/2.

	Volume excavated	Difficulty multiplier	Man hours	Rate per hour	Cost
Max. throw from 1.5 m deep	0.625	1.2	2	£6.50	£9.75
Total volume excavated	0.625			Total cost	£9.75
Dividing total cost by total volume = cost/m^3					£15.60
Profit and oncost 20%					£3.12
Rate per m^3					£18.72

Example 5.7

Excavate to reduce level, starting at ground level and not exceeding 2 m.	m^3

Assume that we will excavate unit length and unit width at any time. Average depth (1.00 + 1.20)/2.

	Volume excavated	Difficulty multiplier	Man hours	Rate per hour	Cost
Max. throw from 1.5 m deep	1.1	1.25	2	£6.50	£17.88
Total volume excavated	1.1			Total cost	£17.88
Dividing total cost by total volume = cost/m^3					£16.25
Profit and oncost 20%					£3.25
Rate per m^3					£19.50

Note that despite the maximum depth being greater than one stage throw, there is in effect no stage throw in reduce level excavation. The result is that the rate for getting out remains the same as the previous item.

The next item, basement excavation, is 'down into' the ground, and throw may have to be taken into account if this proceeds beyond 1.5 m.

Example 5.8

Excavate basement, starting at formation level and not exceeding 1 m maximum depth.	m^3

Assume that we will excavate unit length and unit width at any time.

	Volume excavated	Difficulty multiplier	Man hours	Rate per hour	Cost
Max. throw from 1.5 m deep	0.9	1.25	2.2	£6.50	£16.09
Total volume excavated	0.9			Total cost	£16.09
Dividing total cost by total volume = cost/m^3					£17.88
Profit and oncost 20%					£3.58
Rate per m^3					£21.45

This rate assumes that the getting out will increase the hours/m^3, as it will be done on the low side of the excavations and will require the spoil to be moved over to one side of the excavation.

Example 5.9

Excavate basement, starting at formation level and not exceeding 2 m maximum depth.	m^3

Assume that we will excavate unit length and unit width at any time.

	Volume excavated	Difficulty multiplier	Man hours	Rate per hour	Cost
Max. throw from 1.5 m deep	1.5	1.3	2.2	£6.50	£27.89
Max. throw from 2.0 m deep	0.4	1.35	2.2	£6.50	£7.72
Throw one stage 0.4 m^3	0.4	1	1.4	£6.50	£3.64
Total volume excavated	1.9			Total cost	£39.25
Dividing total cost by total volume = cost/m^3					£20.66
Profit and oncost 20%					£4.13
Rate per m^3					£24.79

Reference to Figure 5.3 will show that although the depths are all within the 1.5 m throw stage, the fact that a deeper separate excavation exists within the shallower one makes a throw a necessity. The volume thrown is not included in total volume.

Example 5.10

Excavate trench starting at bottom of basement excavation and not exceeding 1 m maximum depth	m^3

Without a definitive starting level, this excavation could start at one of two maximum depths below formation level 1 m or 2 m. However, reference to Figure 5.3 will show that the trenches start at the bottom of the more shallow basement excavation. A throw still has to be added, as the initial spoil will be excavated onto the bottom of the basement excavation.

Assume that we will excavate unit length and unit width at any time.

	Volume excavated	Difficulty multiplier	Man hours	Rate per hour	Cost
Max. throw from 1.5 m deep	1	1.45	2.2	£6.50	£20.74
Throw one stage 1 m^2	–	1	1.4	£6.50	£9.10
Total volume excavated	1			Total cost	£29.84
Dividing total cost by total volume = cost/m^3					£29.84
Profit and oncost 20%					£5.97
Rate per m^3					£35.80

Machine excavation

The ubiquitous tractor with backhoe is not the only machine available. Giant earth moving and excavating equipment is capable of moving mountains. Miniature machines can be driven through a 1 m gap to give machine capabilities in back gardens or even inside buildings themselves.

Here we must repeat the earlier caveat regarding the suitability of a particular machine for the type of excavation being tackled. Texts on contractors' plant treat the subject of choice of appropriate equipment in some depth.

With regard to soil conditions, machine excavation is not subject to the same variation in output as hand work, providing the correct type of plant can be procured. The output multipliers in Table 5.5 give some guide as to variations in performance in different soil conditions.

Table 5.5 *Multipliers for difficulty of machine excavation due to nature of ground*

Soil type	Multiplier (× output per hour/m^3)
Vegetable soil, sand, ordinary soil	1.00
Heavy clay	1.33
Soft rock	2.00
Hard rock	5.00*

* Low output due to general inability of machines to do the breaking up which requires pneumatic drills or explosives.

Table 5.6 *Outputs for a tractor with front shovel (1.24 m³) and backhoe (0.06–0.28 m³): loading of spoil is straight into dumpers or tipping lorries alongside excavation*

Excavation	Bucket size (m^3)	Loose granular (small) (m^3/hr)	(hr/m^3)	Compact organic soil (m^3 hr)	(hr/m^3)	Firm cohesive and compact: Soft cohesive (m^3/hr)	(hr/m^3)	Firm cohesive and compact: granular (large) (m^3/hr)	(hr/m^3)
Trenches	0.06	2.5	0.40	2	0.50	1.6	0.63	1.4	0.71
	0.12	4	0.25	3	0.33	2	0.50	1.8	0.56
	0.19	6	0.17	5	0.20	4	0.25	4	0.25
	0.28	9	0.11	8	0.13	7	0.14	6	0.17
Basements	0.06	4	0.25	3.5	0.29	2.5	0.40	2.5	0.40
	0.12	6	0.17	5	0.20	4	0.25	4	0.25
	0.19	9	0.11	8	0.13	6	0.17	6	0.17
	0.28	13	0.08	12	0.08	10	0.10	9	0.11

Table 5.6 shows outputs for a particular machine for different work and conditions. The soil classifications given in the table, which clearly influence the output of the machine, seem to obviate the necessity to use the difficulty multipliers from Table 5.5. This is only partly true. The differences across Table 5.6 are more to do with how much of the spoil is retained in the bucket at each cycle rather than how the machine can cope with penetrating the soil. Unless the work is in soft or hard rock, penetration is not a great problem for the larger modern excavator with good reserves of power. However, the data in Table 5.5 can be useful where excavation rates for ordinary soil only are available.

Examples 5.11 to 5.16 refer again to Figure 5.3. The difficulty multipliers are obtained from Table 5.5, and the machine hours per m³ from Table 5.6.

Example 5.11

Excavate to reduce level, starting at ground level and not exceeding 0.25 m maximum depth.	m³

Assume that we will excavate unit length and unit width at any time.

	Volume excavated	Difficulty multiplier	Hours	Machine Rate per hour	Cost
	0.125	1	0.09	£12.00	£0.14
Total volume excavated	0.125			Total cost	£0.14
Dividing total cost by total volume = cost/m³					£1.08
Profit and oncost 20%					£0.22
Rate per m³					£1.30

The output in machine hours per m³ used for the reduce level excavations have been taken as 20 per cent better than the equivalent rate for basement excavation, the reasoning being that the dumpers or tipper lorries will get alongside the machine at the face being dug.

Example 5.12

Excavate to reduce level, starting at ground level and not exceeding 1 m maximum depth.	m^3

Assume that we will excavate unit length and unit width at any time.

	Volume excavated	Difficulty multiplier	Hours	Machine Rate per hour	Cost
	0.625	1	0.09	£12.00	£0.68
Total volume excavated	0.625			Total cost	£0.68
Dividing total cost by total volume = cost/m^3					£1.08
Profit and oncost 20%					£0.22
Rate per m^3					£1.30

Example 5.13

Excavate to reduce level, starting at ground level and not exceeding 2 m maximum depth.	m^3

Assume that we will excavate unit length and unit width at any time.

	Volume excavated	Difficulty variable	Hours	Machine Rate per hour	Cost
	1.11	1	0.09	£12.00	£1.20
Total volume excavated	1.11			Total Cost	£1.20
Dividing total cost by total volume = cost/m^3				£0.95	£1.08
Profit and oncost 20%					£0.22
Rate per m^3					£1.30

Example 5.14

Basement excavation starting at formation level and not exceeding 1 m maximum depth.	m^3

Assume that we will excavate unit length and unit width at any time.

	Volume excavated	Difficulty variable	Hours	Machine Rate per hour	Cost
	0.9	1	0.11	£12.00	£1.19
Total volume excavated	0.9			Total Cost	£1.19
Dividing total cost by total volume = cost/m³					£1.32
Profit and oncost 20%					£0.26
Rate per m³					£1.58

Example 5.15

Basement excavation starting at formation level and not exceeding 2 m maximum depth.	m³

Assume that we will excavate unit length and unit width at any time.

	Volume excavated	Difficulty variable	Hours	Machine Rate per hour	Cost
	1.9	1	0.11	£12.00	£2.51
Total volume excavated	1.9			Total Cost	£2.51
Dividing total cost by total volume = cost/m³					£1.32
Profit and oncost 20%					£0.26
Rate per m³					£1.58

Example 5.16

Excavate trench starting at bottom of basement excavation and not exceeding 1 m maximum depth.	m³

Assume that we will excavate unit length and unit width at any time.

	Volume excavated	Difficulty variable	Hours	Machine Rate per hour	Cost
	1	1	0.17	£12.00	£2.04
Total volume	1			Total cost	£2.04
Dividing total cost by total volume = cost/m³					£2.04
Profit and oncost 20%					£0.41
Rate per m³					£2.45

Ground water

Excavation below ground water level is measured extra over any type of excavation, irrespective of depth.

Where the whole excavation is below ground water level, the estimator will calculate both full rates – excavation above and excavation below – and subtract the former from the latter to obtain the extra over rate.

In other cases the estimator has to decide for himself how much of any one type of excavation will be below water level. For example, complicated excavations might include all classes of excavation basement, trench and pit, all to different depths – and some or all might be below or partly below water level. The quantity surveyor knows but he does not tell! Outputs vary for each class of excavation and for each depth of each class, and they must do so irrespective of whether there is water in the way or not.

The factors influencing the estimator's judgement on excavation below ground water level are as follows:

- Men working in the excavations require protective clothing.
- Outputs for hand work are generally halved.
- Outputs for machine work are generally halved.
- The proportion of various classes of excavation below water table must be considered.
- Pumping, although given separately in the bill, must be realistically assessed and assumed to give optimal working conditions.

Services

Excavation adjacent to, across or under existing services is also measured extra over any type of excavation .

There may be underground or overground services such as water mains, oil pipe lines, gas pipes and electrical cables, as well as water pipes, telephone wires, district heating mains, culverts, sewers and drain pipes. All of these present a hazard. They may also be old, corroded or already damaged in some way and therefore in a delicate state.

A visit to the offices of the various utilities will determine whether any services are buried on the site. Utilities personnel are willing to spend time with detectors, etc. on site to ensure that you avoid their particular service pipe or cable. A further alternative is to use something like the free-phone services provided by many Regional Authorities to obtain the information.

The factors affecting the estimator's judgement on excavation near services are as follows:

- Some utilities, such as gas and electricity, will require close liaison and this can slow down the work.
- Machine use is restricted at crossings of services.
- Only hand excavation may be allowed where services are adjacent to the excavation.
- Damage to services may result in heavy penalties as well as costs for repairs or replacement.

Example 5.17

Extra over any excavation for excavating below ground water level.	m^3

We will assume that this occurs in the building shown in Figure 5.3, and affects the basement excavation to 2 m deep. The ordinary rate is given in Example 5.9 as £23.53 per m^3 exclusive of profit and oncost.

Assume that we will excavate unit length and unit width at any time.

	Volume excavated	Difficulty multiplier	Man hours	Rate per hour	Cost
To 1.5 m deep	1.5	2.6	2.2	£6.50	£55.77
To 2.0 m deep	0.5	2.7	2.2	£6.50	£19.31
Throw one stage 0.5 m^3	0.5	1	1.4	£6.50	£4.55
Total volume excavated	2			Total cost	£79.63
Dividing total cost by total volume = cost/m^3					£39.81
Deduct basic cost as Example 5.9					£20.66
					£19.16
Profit and oncost 20%					£3.83
Extra over rate per m^3					£22.99

Breaking out

This is the term applied to the removal of material from the excavations which requires the use of equipment other than pick and shovel. These materials are defined in SMM7 as:

- Rock
- Concrete
- Reinforced concrete
- Brickwork, blockwork or stonework
- Coated macadam or asphalt.

It is important to note that it is relatively easier to break out any of these materials if they have been laid down in a thin layer say up to 300 mm thick. It is possible to come across rock in such layers, although there are usually several such layers interleaved with softer material; nevertheless, these are easier to remove. Hand tools such as pick and shovel are of little use against concrete, high quality brickwork or bituminous macadam. A compressor and pneumatic drills are required to break up these materials, whether in bulk or in layers. The hire cost of a compressor and drills or even of electrically powered hammer drills is so low that wedges and hammers are falling out of use.

Once broken up, the material may be loaded for disposal with either a hand shovel or a machine. Outputs and multipliers for excavation in earlier tables reflect the use of appropriate mechanical means of breaking out difficult materials. Generally this means using a compressor with two drills equipped with suitable points. It should be noted that the outputs for hand or machine loading for disposal, while being very different from each other, will not vary much from material to material. In addition, these materials will all bulk up by 33 to 50 per cent. Breaking out with explosives is such a rare occurrence in building works, and so specialized, that it has not been included.

Example 5.18

Extra over any type of excavation for breaking out rock.	m^3

We will assume that there is rock in the trenches in Figure 5.3. The ordinary rate is given in Example 5.10 as £29.84 per m^3 exclusive of profit and oncost.

	Volume excavated	Hours per m^3	Rate per hour	Cost
Compressor and rock points‡	1	7	£4.50	£31.50
	Number			
Labour on compressor and drills	3	7	£6.85	£143.85
	Volume			
Get out loose materials*	1.3	3.5	£6.50	£29.58
Throw one stage*	1.3	1.5	£6.50	£12.68
Total volume excavated	1		Total cost	£217.60
Divide total cost by total volume				£217.60
Deduct basic cost as Example 5.10				£29.84
				£187.77
Profit and oncost 20%				£37.55
Extra over rate per m^3				£225.32

Note
In Example 5.18 * signifies a bulking factor of 30 per cent has been added for getting out and throwing. Note also that the output for the compressor breaking out is very low as the work is being carried out in a trench. In more open conditions the output would be 4 to 4.5 hours/m^3; ‡ signifies that the hourly rate includes fuel, oil, grease, labour in attendance, maintenance and sharpening points.

Example 5.19

Extra over any type of excavation for breaking out a layer of tarmacadam 300 mm thick.	m^3

We will assume that there is an area of old tarmacadam 300 mm thick on the surface where the reduce level excavation will be carried out (Figure 5.3). There are three distinct rates for Examples 5.11, 5.12 and 5.13, which makes it difficult to decide which to use as the base rate. This would normally be done on the basis of the proportion of the bill quantities for each item. In this instance we assume that Example 5.6 gives the largest item: here the ordinary rate is given as £12.72 per m^3 exclusive of profit and oncost.

	Volume excavated	Hours per m³	Rate per hour	Cost
Compressor and chisels	0.3	0.33	£6.85	£0.68
	Number			
Labour on drills	3	0.33	£6.85	£6.78
	Volume			
Get out loose materials*	0.4	2.4	£6.50	£6.24
Total volume excavated	0.3		Total cost	£13.70
Divide total cost by total volume				£45.67
Deduct basic cost as Example 5.6				£15.60
				£30.07
Profit and oncost 20%				£6.01
Extra over rate per m³				£36.08

Note
In Example 5.19 * signifies that an allowance for bulking has been added for getting out.

Working space

The measurement for working space in SMM7 is now given in square metres, provided that the face of the work requiring space is less than 600 mm from the face of the excavation. There is no minimum depth below which working space is not measurable; if it is required it must be measured. At first sight this change in SMM7 might appear to be quite radical, but to the estimator it only rationalises what he has always had to do: assess for himself exactly how much working space would be required, whether or not any was given in the bill!

All previous methods of measurement have treated working space as a theoretical allowance, where the distance from the surface worked upon to the face of the excavation was set in accordance with the depth of the particular class of excavation. Invariably all other factors, such as the nature of the soil, were ignored. Some methods of measurement did try to relate some of the difficulty factors to the measurement, but the allowance was still theoretical.

The problem then that the estimator faces is twofold. It is necessary to decide first how much space the workmen will need, and secondly how many external and internal corners there are on the plan! The first problem requires answers to the following:

- To what depth will the work be carried out?
- What kind of work is involved?
- Is the working space partly or wholly below water level and, if partly, by how much?
- Will the backfilling be of any special materials? If so, then disposal of the excavated material will be required.
- How much additional support will the excavations require?

The second problem only occurs if the building is not rectangular in plan. If it is rectangular then there are always four corners to add to the working space excavation, no matter how many indents or cutouts there are. Should the building plan be a polygon with more than four sides then careful note of the internal and external corners should be made, so that the extra excavation, disposal and support will be properly accounted for.

Having sorted these problems out, the estimator has to decide what width of excavation to allow, cube it up, price it, extend a total and divide by the bill quantity in square metres to arrive at a rate. But what about that width? This is another case for the estimator's skill and working knowledge, backed up by meticulously kept records of earlier work. Every job must be assessed on its merits.

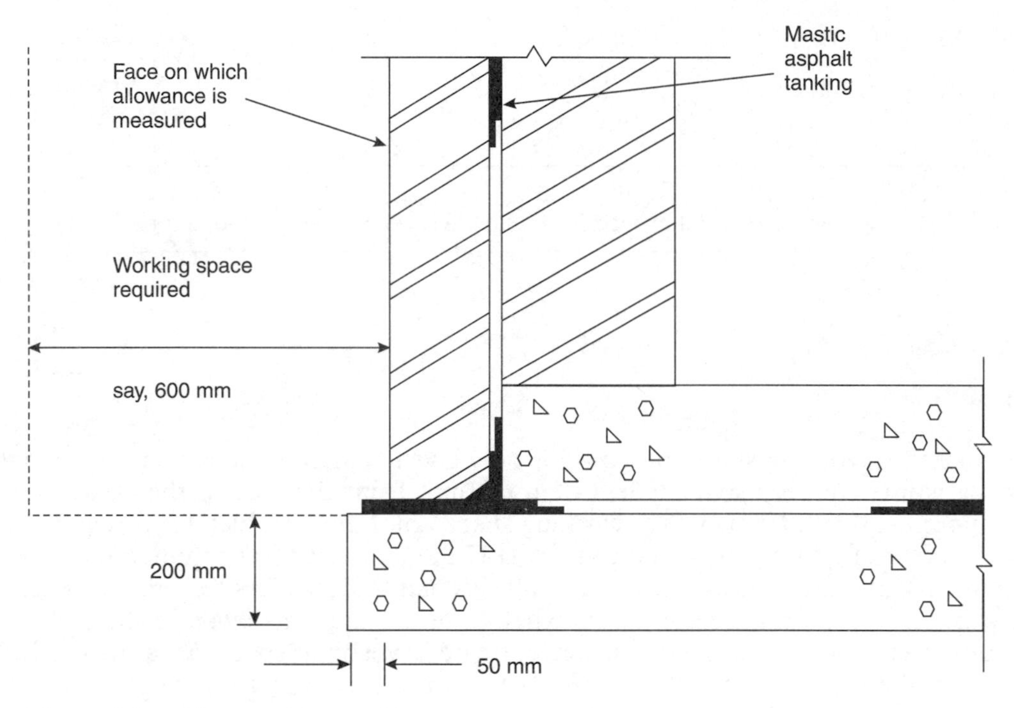

Figure 5.4 *Allowance for working space excavation*

Example 5.20

Extra over basement excavation for working space excavation starting at bottom of shallow basement excavation and not exceeding 1 m deep.	m^2

Refer to Figure 5.4 for details of what causes the need for working allowance, how much is measured and how much we must excavate and backfill. The working space excavation will start below the more shallow basement excavation and go down to the top of the concrete slab. It will therefore be within 1.5 m maximum throw depth.

	Volume excavated	Difficulty multiplier	Man hours	Rate per hour	Cost
Max. throw from 1.5 m deep	1	1.25	2.2	£6.50	£17.88
Backfill	1	1	1.6	£6.50	£10.40
Dispose of surplus*	0.2	1	0.7	£6.50	£0.91
Total volume	1			Total cost	£29.19

Girth of face of protective brickwork	22.00 m
Actual depth of working space excavation	0.80 m
Area for working space given in bill	37.40 m^2
Width for working space	0.60 m
Volume of digging 22.00 × 0.80 × 0.55	9.68 m^3
Plus corners 4 × 0.55 × 0.55 × 0.8	0.97 m^3
Total excavation	10.65 m^3

Excavation cost: total cost × total excavation		£310.76
Additional strutting: 10 reuses, 30 m of 100 mm × 100 mm		£6.60
Timber used	0.3 m^3 at £320.00	£9.60
Labour hours setting strutting per m^3	10 hours at £7.15	£238.33
		£565.30
Profit and oncost 20%		£113.06
		£678.35
Rate per m^2 of working space		£18.14

Assume that we will excavate unit length and unit width at any time.

Note
In Example 5.20 * signifies that bulking must be disposed of.

Earthwork support

Earthwork support is measurable to all faces of excavation except the following:

- Faces not exceeding 250 mm high.
- Sloping faces where the angle of inclination is less than 45° to the horizontal.
- Faces which abut existing structures.

In SMM7 measurement is classed in depth stages and in distances between opposing faces. It appears that the intention is to give gross areas of support where the total depth of excavation falls within any depth stage.

Up to 4 m between opposing faces it is possible to span across with struts of various kinds, e.g. timbers wedged in or proprietary telescopic struts of steel. Over 4 m it is usual to set up raking supports to the lower level of the excavation, although if this is below water level it may be better to use steel work for the shoring, and this will add considerably to the cost.

Other special conditions must be given in the bill, i.e. where the face is curved, below water level (there does not appear to be any provision for support partly below water level), next to roadways, next to existing buildings or in unstable ground, and finally where the support is left in place. The estimator is left to decide how to actually achieve the support, and he will take into account the following additional considerations:

- The most suitable material must be chosen, for example: solid timbers for walings, poling boards or strutting; steel sheet, plywood, etc. for support; proprietary struts; spacing of all members considering the soil type; proprietary close sheeting system.
- The number of uses must be estimated, i.e. the number of times each component can be reused before being discarded. Figures are quoted from 5 to 15 uses.
- Men working in the excavations require protective clothing.
- Outputs for excavation round timbering are generally halved.
- The proportion of various classes of excavation below the water table must be considered.
- Pumping, although given separately in the bill, must be realistically assessed.

Example 5.21

Earthwork support not exceeding 1 m maximum depth and distance between opposing faces not exceeding 2 m.	m^2

The price for this will vary according to the nature of the soil to be supported. For fairly firm, stable soil allow poling boards at 400 to 600 centres, two walings in the height and double strutting at about 2 m centres.

Take a 12 m length of trench and cost it out.

Polings: 2 sides, 12 m long, set at 450 mm centres:			
nr 53 + 2 = 55, each 1.00 × 0.20 × 0.038 m	0.418		
Walings: 2 sides, 2 per side:			
nr 4, each 12 × 0.2 × 0.038 m	0.3648		
Strutting: at 2 m centres gives 6 + 1 = 7 pairs:			
nr 14, each 2 × 0.1 × m	0.28		
Volume of timber	1.0628	£320.00	£340.10
Waste 10%			£34.01
Total Cost			£374.11
Cost per use if 10 uses			£37.41
Labour to unload, set up, strike on completion:			
$1.06\,m^3$ at 10 hours/m^3 at £7.15			£75.99
Cost for trench			£113.40
Trench 12 m long and 1 m deep requires $24\,m^2$ of earthworks support:			
Cost for $1\,m^2$			£4.73
Profit and oncost 20%			£0.95
Rate per m^2			£5.67

Example 5.22

For soil such as we proposed for our imaginary building in Figure 5.3 it might be necessary to put the poling boards closer and use stronger walings than in Example 5.21. This example illustrates the differences.

Take a 12 m length of trench and cost it out.		
Polings: 2 sides, 12 m long, set at 400 mm centres:		
nr $60 + 2 = 62$, each $1 \times 0.2 \times 0.038$ m	0.4712	
Walings: 2 sides, 2 per side:		
nr 4, each $12 \times 0.2 \times 0.05$ m	0.48	
Strutting: at 2 m centres gives $6 + 1 = 7$ pairs:		
nr 14, each $2 \times 0.1 \times 0.125$ m	0.35	
Volume of timber	1.3012 m^3	
Cost at £320.00 per m^3		£416.38
Waste 10%		£41.64
Total Cost		£458.02
Cost per use if 10 uses		£45.80
Labour to unload, set up, strike: 1.30 m^3 at 10 hours/m^3 at £5.59 at £7.15		£93.04
Cost for trench		£138.84
Trench 12 m long and 1 m deep requires 24 m^2 of earthworks support:		
Cost for 1 m^2		£5.78
Profit and oncost 20%		£1.16
Rate per m^2		£6.94

Example 5.23

Earthwork support not exceeding 2 m maximum depth, and distance between opposing faces not exceeding 2 m. m^2

Assume the same loose soil as in Example 5.22.		
Take a 12 m length of trench and cost it out.		
Polings: 2 sides, 12 m long, set at 400 mm centres:		
nr $60 + 2 = 62$, each $2 \times 0.2 \times 0.05$ m	1.24	
Walings: 2 sides, 3 per side: nr 6, each $12 \times 0.2 \times 0.05$ m	0.72	
Strutting: at 1.5 m centres gives $8 + 1 = 9$ sets of 3:		
nr 27, each $2 \times 0.15 \times 0.15$ m	1.215	
Volume of timber	3.175 m^3	
Cost at £320.00 per m^3		£697.40
Waste 10%		£69.74
Total Cost		£767.40
Cost per use if 10 uses		£76.71

Labour to unload, set up, strike: 3.17 m^3 at 10 hours/m^3 at £7.15	£177.20
Cost for trench	£253.91
Trench 12 m long and 2 m deep requires 48 m^2 of earthwork support:	
Cost for 1 m^2	£5.29
Profit and oncost 20%	£1.06
Rate per m^2	£6.35

Disposal of water

SMM7 recognizes two sources of water affecting excavations: water present in the ground, and water accumulating on the surface of the ground. The estimator is expected to give lump sum prices for dealing with either or both. The former arises when excavation takes place below the natural water table of the ground and the water wells up into the excavation. The lump sum price given in the bill by the estimator is adjusted in the final account in relation to measurements of ground water levels taken for each excavation as they are done.

The following points are valid for both classes of disposal:

- Soil type.
- Topography of the site.
- Methodology adopted by the contractor for the construction in general and for the disposal of ground and or surface water in particular.
- Time of year and expected weather pattern.
- Records from previous contracts in the area.

Disposal of excavated material

The primary factors in disposal of excavated material are as follows:

- What type and condition is the material?
- Is it suitable for backfilling?
- Is it suitable for general filling?
- Has it to be disposed of in a particular way, i.e. in spoil heaps for later use, or to a coup?
- Does the contractor have a choice?

Other factors are the distance to spoil heaps or the proximity of a tip external to the site, the value of any surplus material, and whether or not the client wishes to retain a proprietary interest in it up to the point of disposal. For example, hand excavation followed by disposal into spoil heaps on site carried out using wheelbarrows might be reasonable if the wheeling distance was not too far, the ground was suitable, or barrow runs could be economically provided. A maximum distance might be in the region of 150 m. Beyond this distance mechanical barrows or small dump trucks might be more suitable, bearing in mind that mechanical devices run on four wheels and need wider roadways, and that they bog down in poor soil conditions.

Wheeled or tracked plant engaged in bulk excavations such as reduced level or basement work could 'wheel' each bucket load to a disposal point, but this is only economic if the bucket is of reasonable capacity, the wheeling distance is less than 50 m and the machine can get out of the excavation to deposit the load. For distances beyond this it is considered more economical to utilize dump trucks or light tramway and hutches for on-site disposal, and tipping lorries for off-site disposal.

When considering disposal, the estimator must be aware that excavated material is measured in the bill as the volume of the void to be created, i.e. the volume of compacted spoil removed. As soon as it is removed it will increase in volume. This phenomenon is called bulking and is ignored by the quantity surveyor when measuring disposal. However, the estimator must allow in his rates for the fact that whenever he disposes of one metre cube of spoil as measured, the men on site will be handling one and a bit metres cube of spoil (see Table 5.7).

The corollary of course is that the cubic metre handled on site is not as heavy as when it was in the ground. This is important when loading lorries which have axle weight restrictions when on the public roads. For example, loading a 6-tonne payload lorry with soil having a weight of $1.5\,t/m^3$ and a bulking rate of 15 per cent means that the lorry might carry $(6/1.5) + 15$ per cent $= 4.6\,m^3$. However, the soil must not be beyond the bulk capacity of the lorry body, causing it to overflow onto the roads; the authorities do not allow that either.

Matching the output of the men and/or machines digging to the means of disposal is of great importance if maximum efficiency is to be achieved. Whether disposal is by barrow or lorry, it is costly to have the excavation lagging behind the disposal or vice versa. They must match, as explained in Chapter 3 on mechanical plant.

In October 1996 HM Government introduced a 'Landfill Tax'. The aim is to encourage the recycling of building materials, a reduction in the amount of material dumped in landfill sites and encouragement of more responsible disposal of hazardous waste. The tax is administered by HM Customs and Excise, an agency with the broadest and strictest powers of all government agencies within the United Kingdom. The level of taxation was introduced on a sliding scale, increasing each year to a ceiling of £15.00 per tonne by April 2004. There are two levels of tax, the lower one for 'inert' waste and the higher one for 'active waste'. As with all legislation the whole thing is bound by 'ifs' and 'buts'. Greater detail can be obtained by visiting the web site at http://www.hmce.gov.uk and searching there for Landfill Tax. Payment of the tax on a construction contract can be tackled in two ways. The estimator can include the amount in each bill item priced or a sum/per cent can be included in the Preliminary Works Bill. In very rare cases, the Building Employer will undertake to pay the tax but a sum would be included in the Bill of Quantities (probably as a Provisional Sum) so that the cost was included in any budgeting. The tax is NOT taken into account for the examples in this chapter as at the time of writing there is some doubt about the classification of the waste as inert or active and without quantities it is not entirely possible to ascertain the correct amount of tax without being able to calculate mass.

Table 5.8 summarizes the factors for consideration in excavation and disposal.

Table 5.7 *Range of increase in hulk for various soil types*

Type of soil	Rate of bulking (%)
Vegetable soil	25–30
Gravel	10–20
Sand	10–12.5
Clay	25–50
Chalk	33–50
Rock	40–50

Table 5.8 *Summary of factors affecting rates and choice of hand or machine excavation and disposal*

Type of excavation

- Shallow area: vegetable soil excavation
- Open excavation: reduced level, basement
- Confined excavation: trench
- Isolated excavation: pit

Nature of excavation

- New site: unrestrained or no access for machines
- Existing site or buildings: no access for machines; inside existing building
- Restrictive conditions by client or local authority
- Work restricted to certain hours
- Routes to tip restricted
- Compressors not allowed
- Blasting not allowed

Nature of ground

- Dry, wet, frozen
- Clay, sand gravel, rock etc: degree of compaction, requiring pick and shovel or compressor and tools; rate of bulking
- Water table: above or below; seasonal fluctuations; tidal fluctuations; pumping required

Weather

- Frozen ground as above
- Diversion or pumping of surface water

Volume of excavation

- Too large for just hand excavation
- Too little to warrant expense of machine
- Mixed: bulk by machine, trim by hand, small isolated by hand

Variety of excavation

- Requires more than one type of machine

Disposal

- Haul to tip: fees; spreading at tip; length of haul
- On site: spread and level; spoil heaps re-excavated; dispose, distance of wheel

Example 5.24

Sundry items of disposal by hand. Barrows and planking for barrow runs are included in general overhead percentages.

Wheel 50 m and deposit in heaps.	m^3

		Labourer rate	
Load into barrow	1.5		
Wheel and deposit	0.7		
Total hours	2.2	£6.50	£14.30
Profit and oncost 20%			£2.86
Rate per m^3			£13.99

+100 m and deposit in heaps.	m^3

		Labourer rate	
Load into barrow	1.5		
Wheel and deposit	0.7		
Additional 50 m	0.7		
Total hours	2.9	£6.50	£18.85
Profit and oncost 20%			£3.07
Rate per m^3			£18.44

Wheel 50 m and deposit, spread and level.	m^3

		Labourer rate	
Load into barrow	1.5		
Wheel and deposit	0.7		
Spread and level	0.4		
Total hours	2.6	£6.50	£16.90
Profit and oncost 20%			£3.38
Rate per m^3			£20.28

Return, fill and ram with material from the excavations.	m^3

		Labourer rate	
Excavate from spoil heap	1.5		
Return, fill and ram	1.6		
Total hours	3.1	£6.50	£20.15
Profit and oncost 20%			£4.03
Rate per m^3			£24.18

Note that in the last example the spoil is assumed to be at the side of the excavation where it was left immediately on excavation.

Example 5.25

A basement excavation is being dug by tracked excavator. The excavator and the dumpers will gain access to the excavations by a temporary earth ramp. We will assume that the excavation has been rated and we now wish to price the disposal to a coup two miles from the site. There is a total of 740 m^3 of excavation. Calculate the rate per m^3.

The excavator has a 0.57 m^3 bucket and is expected to operate in the region of 20 cycles per hour. Each cycle is: excavate one full bucket; off-load into dump truck; return to face of excavations. This gives an excavation rate of 11.4 m^3 per hour. The excavator will be operative for 35 hours per week.

The material is assumed to bulk up by 15 per cent on excavation to 13.11 m^3. Dump trucks must therefore be capable of handling 13.11 m^3 per hour. The excavation rate produces 1 m^3 in 60/13.11 = 4.58 minutes. The time taken to load a 2 m^3 (3 tonne) dump truck is then 9.15 minutes. The time taken to reach the tip at 20 mph is 60(2/20) = 6 minutes, and the return time is another 6 minutes. The time taken to tip the load is say 3 minutes. Thus:

$$\text{number dumpers required} = \frac{\text{time to load and haul}}{\text{time to load}} = \frac{9.15 + 15}{9.15} = 2.639$$

This is rounded up to three trucks. It is better to have a truck idle for a short period than to have a more expensive excavator sitting around waiting for every dump truck to return. Besides which, it is perfectly possible for dump trucks to be delayed in traffic at peak periods in the day.

Hire of dump trucks and road licence per week			£160.00
Driver for 39 hour week at £6.85			£267.15
			£427.15
Cost per hour over 35 hour week			£12.20
Cleaning and maintenance at half hour per day of 8 hours, i.e. 1/16 of driver cost per hour			£0.43
	Consumption	Cost	
Fuel	2.5	£0.90	£2.25
Oil and grease, say, per hour			£0.20
Insurance at 5% of cost per hour			£0.61
Cost of one dumper truck			£15.69
Cost of two more dumper trucks			£31.39
Cost per hour			£47.08
The output is related to the excavator at 13.11 m^3/hour.			
Divide cost/hour of dump trucks by excavator output/hour			£3.59
Transport to site, own wheels, say half hour each way		£47.08	
Divide by expected total output of		740 m^3	£0.06
Local dumping charge per load of 2 m^3*		£12.00	
charge per m^3		£6.00	£6.00
			£9.65
Profit and oncost 20%			£1.93
Rate per m^3			£11.59

* Exclusive of Landfill Tax

Filling to excavations

The volume of fill given in the bill should be the volume of the void to be filled. Thus, whether the material used is to come from the excavation in the first instance or to be imported to the site, the estimator must be aware that the volume of material required is much greater, for it will require compaction after placing. For example, hardcore fill decreases by 25 per cent of its original bulk after placing and compaction.

SMM7 gives two classifications of filling: filling to excavations and filling to make up levels. Filling to excavations is generally more labour intensive or more difficult by machine since it involves backfilling around the actual building works. Filling to make up levels is assumed to mean bulk full over more extensive areas either inside the building or in the open, where it would be possible to load in with a machine or by directly tipping a truck. The latter is obviously the cheapest solution, although not strictly speaking the best construction method since any fill should be laid down in layers and compacted before the next layer is superimposed.

SMM7 also gives two thickness classifications, up to 250 mm and more than 250 mm, thus recognizing the additional labour in spreading thinner layers over wide areas.

Example 5.26

Filling exceeding 250 mm thick to make up levels with material from the excavations, well rammed in layers of 250 mm thickness.	m^3

Excavation in spoil heaps	1.80		
Barrow in and fill	0.80		
Mechanical punning	0.50		
Total hours	3.10	£6.50	£20.15
Mechanical punner hire inc. fuel, oil etc.: 0.5 hours at		£3.00	£1.50
			£21.65
Profit and oncost 20%			£4.33
Rate per m^3			£25.98

Example 5.27

Filling not exceeding 250 mm average thickness to make up levels with hardcore of broken brick or stone, blinded with ashes.	m^3

From inspection of the drawings we ascertain the hardcore bed is 150 mm thick. Allow 6 m^2 per cubic metre.

Labour and Plant:	
Barrow in and fill hardcore 1 m^3: 1.3 hours at £6.50	£8.45
Mechanical punning: 6 × 0.13 hours at £6.50	£5.07
Spread and consolidate blinding: 6 × 0.7 hours at £6.50	£27.30
Punner hire inc. fuel, oil etc.: 0.78 hours at £3.00	£2.34
	£43.16

Material:	
Hardcore, 1.6 tonne/m^3: per m^3 at £8.00 tonne	£12.80
20% loss in compaction, add 25%	£3.20
Ash blinding, 0.8 tonne/m^3: per m^3 at £9.00 tonne	
Layer 25 mm thick over 6 m^2 gives 0.15 m^3: gives 0.15 tonnes	
Therefore cost per m^3 of hardcore is	£1.35
	£17.35
Profit and oncost 20%	£3.47
Rate per m^3	£20.82

Surface packing

According to SMM7, no item need be given for achieving a finished surface anywhere between the horizontal and a slope of 15°. Over 15° there are two classifications: battered and vertical. Battered surfaces up to the natural angle of repose of the material can be readily achieved using machines, but beyond this angle material would have to be packed or built by hand. The angle is usually in the region of 40°. Packing by hand is very time consuming.

Surface treatments

Five categories are given in SMM7:

1. Application of herbicides is self-explanatory, but it should be pointed out that men applying herbicides do require protective clothing, proper washing facilities and the means to dispose of or safely store protective clothing.
2. Compacting is subdivided into three classifications and needs no explanation.
3. Trimming the sides of cuttings or embankments is subdivided into battered, vertical or rock; the last would be natural rock and would therefore only occur in cuttings.
4. Trimming rock to a fair or exposed face.
5. Preparing subsoil for topsoil.

Some features of these have already been covered in the examples.

Example 5.28

Grade bottom of excavation to a level surface	m^2

Grade/level: 0.08 hours at £5.30	£6.50	£0.52
Profit and oncost 20%		£0.10
Rate per m^2		£0.62

Example 5.29

Grade and compact surface of hardcore bed to a slope of 20°	m^2

Hand pack hardcore: 0.13 hours at £6.50	£0.85
Profit and oncost 20%	£0.17
Rate per m^2	£1.01

6

Concrete work

SMM7 Sections E11, E20, E30, F31

In situ concrete

Concrete is a mixture of cement, fine aggregate, coarse aggregate and water. The strength and durability of concrete are dependent on the proportions of materials in a particular mix. The methods for specifying prescribed and designed mixes of concrete, either site mixed or ready mixed, are covered by BS 5328.

Designed mixes

Designed mixes are those for which the purchaser is responsible for specifying the performance, and the producer is responsible for selecting mix proportions that will conform with the particular performance requirements.

Prescribed mixes

Prescribed mixes are those for which the purchaser specifies the mix proportions that will produce a concrete with the required performance. There are two types of prescribed mix:

Ordinary prescribed mixes

These are specified in accordance with the requirements of BS 5328. The grade of concrete is prescribed in Table 1 of the standard, together with permitted types of cement and aggregate, aggregate size, etc.

Special prescribed mixes

These generally meet the same requirements as ordinary prescribed mixes, but mix proportions in kilograms are given for each constituent material.

Materials

Cement

The most widely used cement is Ordinary Portland Cement, which is manufactured to perform to B2 12. In the examples that follow we use ordinary Portland cement. There are of course other types of cements produced to satisfy particular criteria.

Aggregate

The term 'aggregate' is used to describe the gravels, crushed stone and other materials which are mixed with cement and water to make concrete. Aggregates defined as fine aggregate are natural sands, crushed rock or gravel or other materials which will pass through a 5 mm BS sieve. Coarse aggregate, such as natural gravel, crushed rock or crushed gravel, is retained in this sieve.

Water

Water for use in concrete should be reasonably free from such impurities as suspended solids, organic matter and dissolved salts, which may adversely affect the properties of the concrete.

Admixtures

No admixtures are permitted to be incorporated in ordinary prescribed mixes. Therefore the development of costs in the examples in this chapter does not include costs for admixtures. However, the estimator must be aware of and account for any such requirement in the specification for admixtures in designed or special prescribed mixes.

Weights and volumes

The unit of measurement for the main structural items of *in situ* concrete work is the cubic metre (m^3). However, constituent materials that form concrete are sold by weight. Thus we must develop the costs for such items by weight, and then convert the costs by weight to costs by volume for dry materials. Table 6.1 gives average weights per cubic metre for various materials.

Weights of coarse and fine aggregates may vary for a number of reasons, and the estimator can account for these variations where required. For example, the weight per volume depends on how wet or dry the material is, and on the source of the material.

The mass of aggregate to be used with 100 kg of cement can be obtained from Tables 1 and 2 of BS 5328.

Shrinkage in concrete mix

During the process of mixing materials to form concrete, a reduction in bulk takes place. This is due to the finer particles of sand and cement filling the voids or interstices of the

Table 6.1 *Average weight per cubic metre of various materials*

Material	Tonnes/m^3
Cement in bags	1.28
Cement in bulk	1.28 to 1.44
Wet sand	1.60
Damp sand	1.28
Aggregate	1.60

BS 5328 Table 1. Mass of dry aggregate to be used with 100 kg of cement

Grade of concrete	Nominal maximum size of aggregate (mm)	40		20		14		10	
	Workability	Medium	High	Medium	High	Medium	High	Medium	High
	Range for standard sample (mm)	50–100	80–170	25–75	65–135	5–55	50–100	0–45	15–65
	Range for sample taken in accordance with 9.2 (mm)	40–110	70–180	15–85	55–145	0–65	40–110	0–55	5–75
		kg	kg	kg	kg	kg	kg	kg	kg
C7.5P	Total aggregate	1080	920	900	780	N/A	N/A	N/A	N/A
C10P	Total aggregate	900	800	770	690	N/A	N/A	N/A	N/A
C15P	Total aggregate	790	690	680	580	N/A	N/A	N/A	N/A
C20P	Total aggregate	660	600	600	530	560	470	510	420
C25P	Total aggregate	560	510	510	460	490	410	450	370
C30P	Total aggregate	510	460	460	400	410	360	380	320

N/A not applicable

BS 5328 Table 2. Percentage by mass of fine aggregate total aggregate

Grade of concrete	Nominal maximum size of aggregate (mm)	40		20		14		10	
	Workability	Medium	High	Medium	High	Medium	High	Medium	High
C7.5P, C10P, C15P		30–45		35–50		N/A		N/A	
C20P, C25P, C30P	Grading zone 1	35	40	40	45	45	50	50	55
	Grading zone 2	30	35	35	40	40	45	45	45
	Grading zone 3	30	30	30	35	35	40	40	45
	Grading zone 4	25	25	25	30	30	35	35	40

N/A not applicable

Notes on the use of Tables 1 and 2

NOTE 1. The proportions given in the tables will normally provide concrete of the strength in N/mm^2 indicated by the grade except where poor control is allied with the use of poor materials.
NOTE 2. For grades C7.5P, C10P and C15P a range of fine-aggregate percentages is given; the lower percentage is applicable to finer materials such as zone 4 sand and the higher percentage to coarser materials such as zone 1 sand.
NOTE 3. For all grades, small adjustment in the percentage of fine aggregate may be required depending on the properties of the particular aggregates being used.
NOTE 4. For grades C20P, C25P and C30P, and where high workability is required, it is advisable to check that the percentage of fine aggregate stated will produce satisfactory concrete if the grading of the fine aggregate approaches the courser limits of zone 1 or the finer limits of zone 4.

coarse aggregate. In order to attain the required volume in a particular mix we have to increase the quantities of material to compensate for the reduction in bulk. This is done by calculating the cost based on the mix proportions of the concrete and adding a percentage adjustment to allow for the cost of the additional material. There are various methods of adjusting for shrinkage in concrete, as detailed below. The preference of method will vary from estimator to estimator.

Additional material

In our calculations for the costs of concrete we can anticipate shrinkage of between 20 and 33.33 per cent. Allowances for the costs of additional material are determined as follows.

Shrinkage of 20 per cent

A volume of 1 m^3 less 20 per cent gives 0.80 m^3. Therefore, to obtain 1 m^3 we shall have to add 0.20 m^3. This represents an increase of:

$$\frac{0.20}{0.80} \times 100 = 25 \text{ per cent}$$

Shrinkage of 28.6 per cent

A volume of 1 m^3 less 28.6 per cent gives 0.714 m^3. Therefore to obtain 1 m^3 we shall have to add 0.286 m^3. This represents an increase of

$$\frac{0.286}{0.714} \times 100 = 40 \text{ per cent}$$

Shrinkage of 33.33 per cent

A volume of 1 m^3 less 33.33 per cent give 0.667 m^3. Therefore to obtain 1 m^3 we shall have to add 0.333 m^3. This represents an increase of

$$\frac{0.333}{0.667} \times 100 = 50 \text{ per cent}$$

Costing for shrinkage

Where the lowest percentage for shrinkage is used, it is practice to divide the cost for the concrete mix by the combined volumes of the fine and coarse aggregates. Where the higher percentages are used, it is practice to divide the cost for the concrete mix by the combined volumes of the cement and the fine and coarse aggregates.

Example 6.1

Ordinary prescribed mix; grade C20P concrete, 40 mm aggregate, medium workability.	m^3

From Table 1 of BS 5328 we can see that for every 100 kg of cement we require 660 kg total aggregate. We must therefore refer to Table 2 of BS 5328 to establish the percentage by mass of fine aggregate; for grading zone 2 this is 30 per cent. Therefore the mix proportions are as follows:

Cement	100 kg
Fine aggregate	198 kg
Coarse aggregate	462 kg

The volume of mix based on the use of 1 tonne of cement is calculated as follows:

Cement	1.00 t ÷ 1.28 t/m³	0.78
Fine aggregate	1.98 t ÷ 1.50 t/m³	1.32
Coarse aggregate	4.62 t ÷ 1.60 t/m³	2.89
Total volume of mix		4.99 m³

We develop the costs for a mix based on the use of 1 tonne of cement. We assume that the cement is delivered in 50 kg bags, which are kept in a dry store and will require to be transferred from the store to the mixer. Aggregates, on the other hand, are stored in bulk adjacent to the mixer.

Cement	1.00 t at £90.00/t	£90.00
Fine aggregate	1.98 t at £19.00/t	£37.62
Coarse aggregate	4.62 t at £24.00/t	£110.88
Transfer 1 t cement: labourer 1 hour/t	1.00 hr at £6.50/t	£6.50
Cost of mix		£245.00
Allow shrinkage 40%		£98.00
Material cost of 4.99 m³		£343.00
Material cost per m³		£68.74

Example 6.2

Ordinary prescribed mix; grade C25P concrete, 20 mm aggregate, medium workability.	m³

From Table 1 of BS 5328 we can see that for every 100 kg of cement we require 510 kg total aggregate. We must therefore refer to Table 2 of BS 5328 to establish the percentage by mass of fine aggregate; for grading zone 2 this is 35 per cent. Therefore the mix proportions are as follows:

Cement	100 kg
Fine aggregate	178.5 kg
Coarse aggregate	331.5 kg

The volume of mix based on the use of 1 tonne of cement is calculated as follows:

Cement	1.00 t ÷ 1.28 t/m³	0.78
Fine aggregate	1.79 t ÷ 1.50 t/m³	1.19
Coarse aggregate	3.32 t ÷ 1.60 t/m³	2.08
Total volume of mix		4.05 m³

We develop the costs for a mix based on the use of 1 tonne of cement.

Cement	1.00 t at £90.00/t	£90.00
Fine aggregate	1.79 t at £19.00/t	£34.01
Coarse aggregate	3.32 t at £24.00/t	£79.68

Transfer 1 t cement: labourer 1 hour/t	1.00 hr at £6.50/t	£6.50
Cost of mix		£210.19
Allow shrinkage 40%		£84.08
Material cost of 4.05 m^3		£294.27
Material cost per m^3		£72.66

Mixer costs

Only small quantities of concrete will be mixed on site by hand. Larger quantities will arrive on site ready mixed by an independent producer, or will be mixed on site using a machine. Purchasing concrete from a producer eliminates the requirement to calculate the costs of material and mixing; only the costs for final transporting, placing etc. have to be added to the producer's quoted price.

Where a mixer is used, we must establish its cost. A typical method of assessing the mixer cost per hour is given in Example 3.1. In this chapter, mixer costs are given without calculation.

Labour costs

Having established how to calculate the costs for concrete mixes and mixers, we have now to determine the cost of labour required to produce, transport and place the concrete mix. As the essence of good concrete work is that continuous production be maintained, the type of job to be done and the mixer being used determines the number of men in a gang or squad.

We could adopt labour constants; that is, average outputs for each type of concrete component, which would account for producing, transporting and placing a mix. However, a more accurate way to develop the costs is to determine the output of a mixer in m^3 per hour and to estimate the gang size required to mix, transport and place. The gang cost per m^3 is then calculated by dividing the gang cost per hour by the mixer output in m^3 per hour.

The rate of discharge is determined by the demand on the mixer on site and by its capacity in terms of number of batches per hour.

Mixer outputs and gang sizes

Figure 6.1 shows mixer outputs and related gang sizes. The gangs given on the graph are for mixing only and would be made up as follows:

One-man gang	Mixer operator
Two-man gang	Mixer operator, feeder operator
Three-man gang	Mixer operator, feeder operator and labourer.

It has been assumed that mechanically fed mixers are used, i.e. those with a feed hopper and dragline aggregate feeder. Where a mixer is fed manually, another labourer should be added to the squad.

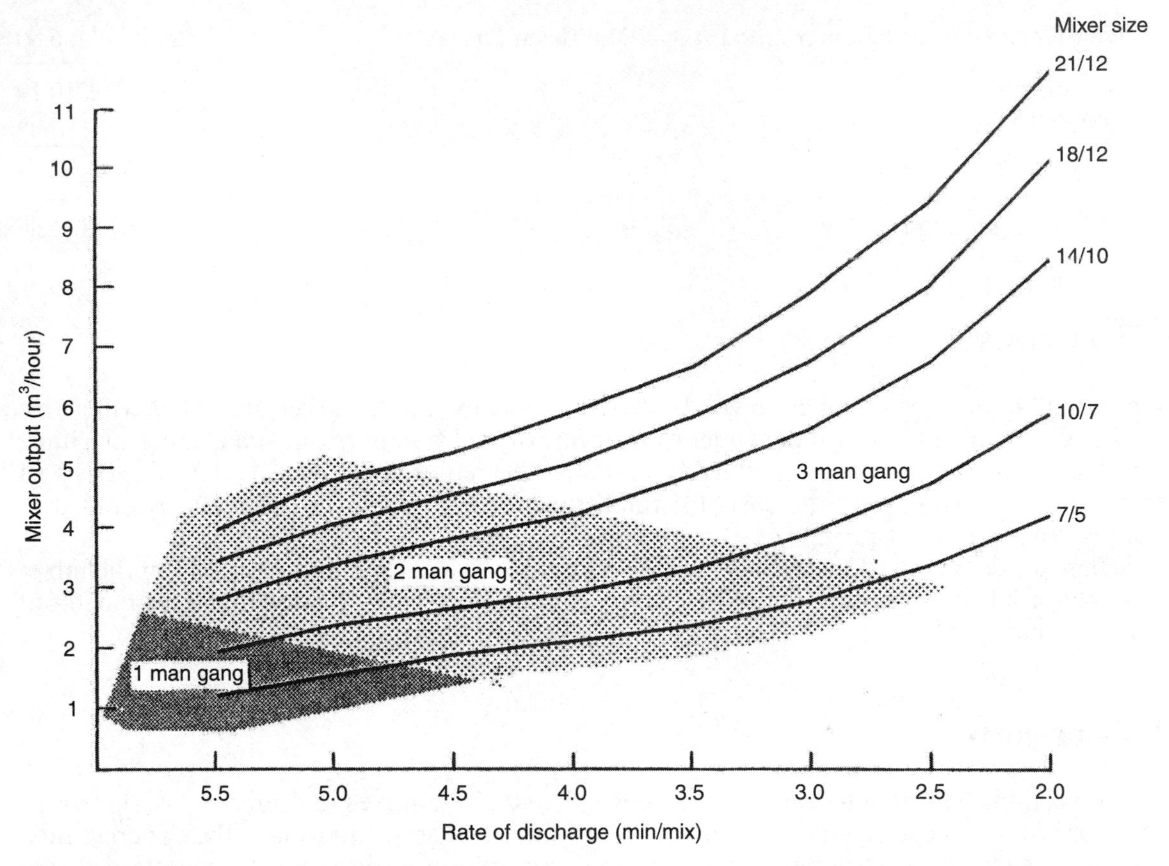

Figure 6.1 *Mixer outputs and related gang sizes*

Under the Working Rules agreement WR3, extra payments are made for extra skill or responsibility. Operatives responsible for a mixer (one per mixer) and for a dragline feeder are entitled to such extra payments, which are given in the Working Rules.

Placing outputs

Placing outputs are dependent on the type of concrete member to be poured and the mixer output. The gang size will therefore depend on these two factors. Table 6.2 indicates required gang sizes for a selection of typical members.

Transporting outputs

Concrete is normally transported from its point of mixing to the required position on site by ordinary wheelbarrow or diesel powered dumper. On the average building site, where the terrain is rough and messy, a wheelbarrow is uneconomical for transporting concrete beyond about 60 m. A dumper can be used on distances of up to 400 m. However, the aggregates in a concrete mix can be settled out by the action of a dumper, and therefore it is better to minimize journey lengths.

A labourer transporting concrete using a wheelbarrow will do so at about 1 m^3/hour over a 100 m round trip.

Table 6.2 *Gang sizes required to pour particular members*

Members and description	Mixer output (m^3/hour) 0–4	4–8	8–12
Foundations in trenches, over 300 mm thick	1	1	2
Foundations in trenches, not exceeding 300 mm thick	1	2	2
Beds not exceeding 150 mm thick	2	2	3
Beds 150–450 mm thick	1	2	2
Beds over 450 mm thick	1	1	2
Walls not exceeding 150 mm thick	2	3	3
Walls 150–450 mm thick	2	2	3
Walls over 450 mm thick	2	2	3
Columns	2	2	3
Staircases	2	3	3

A standard dumper of 368 litres working at 20 per cent below maximum capacity can take full discharge from mixers up to 14/10 capacity. For larger mixers, a swivel skip dumper of say 510 litres would need to be used. Dumper costs can be calculated using the method adopted earlier for mixer costs. An average speed of 30 km/hour can be used in the calculation of dumper transport costs.

Other plant may be needed, such as a hoist to take a concrete mix from ground level to floor levels above. If this is the case then costs for such plant would have to be assessed and included in the rate build-up.

Example 6.3

Plain *in situ* concrete in foundations; grade C25P, 20 mm aggregate, medium workability; poured on or against earth or blinded hardcore; not exceeding 300 mm thick. m^3

We can begin by taking the materials cost of the mix direct from Example 6.2

Materials

Cost per m^3	£72.66
Waste: add 5%	£3.63

Mixer

We will use a 10/7 concrete mixer, and assume that the cost for this mixer is £1.06 per hour. From the mixer output graph (Figure 6.1) we can see that a 10/7 mixer discharging at say 3 minutes per mix will produce approximately 4 m^3/hour. Therefore

Mixer cost per m^3: £1.06/4	£0.27

Mixing squad

From Figure 6.1 we see that the gang size required is 3 men. Therefore

Mixer operator cost/hour	£6.73	
Feeder operator cost/hour	£6.77	
Labourer cost/hour	£6.50	
Gang cost/hour	£20.00	
Gang cost per m^3: £20.00/4		£5.00

Transporting
Transporting mix by wheelbarrow over say a 50 m round trip, assuming that 1 labourer will transport 1 m^3/hour over 100 m:

Labourers (nr) to barrow 4 m^3/hour: $(50/100) \times 4 = 2$	
Cost to transport 4 m^3: $2 \times$ £6.50 = £13.00	
Cost to transport 1 m^3: £13.00/4	£3.25

Placing
From Table 6.2 we see that 2 labourers are required to spread the mix. Therefore

Cost per m^3 for 1 labourer at £6.50: £6.50/4	£1.63
	£86.44
Profit and oncost 20%	£17.29
Rate per m^3	£103.73

Example 6.4

> Plain *in situ* concrete in beds; grade C25P, 20 mm aggregate, medium workability; poured on or against earth or blinded hardcore; 150–450 mm thick. m^3

Again we can take the materials cost of the mix direct from Example 6.2

Materials

Cost per m^3	£72.66
Waste: add 5%	£3.63

Mixer
We will use a 14/10 concrete mixer, and assume that the cost for this mixer is £2.38 per hour. From the mixer output graph (Figure 6.1) we can see that a 14/10 mixer discharging at say 3 minutes per mix will produce approximately 5.75 m^3/hour. Therefore

Mixer cost per m^3: £2.38/5.75	£0.41

Mixing squad
From Figure 6.1 we see that the gang size required is 3 men. Therefore

Mixer operator cost/hour	£6.73
Feeder operator cost/hour	£6.77
Labourer cost/hour	£6.50
Gang cost/hour	£20.00

Gang cost per m^3: £20.00/5.75	£3.48

Transporting
Transporting mix by wheelbarrow over say a 100 m round trip, assuming that 1 labourer will transport 1 m^3/hour over 100 m:

Labourers (nr) to barrow 5.75 m^3/hour: (100/100) × 5.75 = 5.75 (say 6)	
Cost to transport 4 m^3: 6 × £6.50 = £39.00	
Cost to transport 1 m^3: £39.00/6	£6.50

Placing
From Table 6.2 we see that 2 labourers are required to spread the mix. Therefore

Cost per m^3 for 2 labourer at £6.50: 2 × £6.50/5.75	£2.26
	£88.94
Profit and oncost 20%	£17.79
Rate per m^3	£106.73

Example 6.5

Reinforced concrete walls; grade C25P, 20 mm aggregate, medium workability; not exceeding 150 mm thick.	m^3

The following costs per m^3 are taken from Example 6.4:

Materials	£72.66
Waste: add 5%	£3.63
Mixer	£0.41
Mixing squad	£3.48
Transporting	£6.50

Placing costs are now added. From Table 6.2 we see that 3 labourers are required to spread the mix. Therefore

Cost per m^3 for 3 labourers at £6.50: 3 × £6.50/5.75	£3.39
	£90.07
Profit and oncost 20%	£18.01
Rate per m^3	£108.08

Reinforcement for *in situ* concrete

Bar reinforcement

Reinforcement bars are generally delivered to site cut and bent in accordance with reinforcement drawings and bending schedules. Bars are fixed and secured in position using

Table 6.3 *Average quantities of tie wire for reinforcement bar*

Reinforcement bar diameters (mm)	Tie wire per 1000 kg of bar, inc. waste (kg)
6–12	12
16–25	7
32–50	5

Table 6.4 *Labour constants for cutting and bending bar reinforcement (steel fixer hours per 1000 kg of steel bar)*

	Reinforcement bar diameter (mm)									
Concrete member	6	8	10	12	16	20	25	32	40	50
Foundations and beds	50	42	35	30	28	25	23	–	–	–
Slabs	52	42	35	30	28	25	23	–	–	–
Walls	50	42	35	30	28	25	23	–	–	–
Beams, columns and staircases	55	46	39	33	30	27	25	25	27	27
Beam and column casings	50	42	35	30	28	–	–	–	–	–
Links and stirrups	80	70	60	55	50	45	–	–	–	–

Table 6.5 *Labour constants for fixing bar reinforcement (steel fixer hours per 1000 kg of steel bar)*

	Reinforcement bar diameter (mm)									
Concrete member	6	8	10	12	16	20	25	32	40	50
Foundations and beds	50	42	35	30	28	25	23	–	–	–
Slabs	52	44	37	32	30	28	27	–	–	–
Walls	70	60	50	45	40	35	30	–	–	–
Beams, columns and staircases	70	60	50	45	40	35	30	28	26	24
Beam and column casings	60	58	42	36	33	–	–	–	–	–
Links and stirrups	90	80	70	55	50	45	–	–	–	–

spacers and tie wire. An allowance for spacers is made in the examples that follow. Average quantities for tie wire per 1000 kg of reinforcement bar are given in Table 6.3.

Where cutting, bending and fixing bars is carried out on site, 5 per cent wastage should be added. On the other hand, where reinforcement is delivered to site cut and bent, 1 per cent should be added to allow for damage and loss.

Qualified bar benders and fixers are paid the craft operatives' wage in accordance with the Working Rules.

Table 6.4 gives labour constants for cutting and bending. Table 6.5 gives labour constants for fixing.

Example 6.6

12 mm diameter hot rolled plain round mild steel bars in foundations, straight or bent.	tonne

Materials	
1000 kg of 12 mm bars delivered cut and bent	£340.00
Waste: add 1%	£3.40
	£343.40
Unload and stack: labourer 1 hour × £6.50	£6.50
	£349.90
Tie wire: 12 kg × £17.80/kg	£213.60
Chairs and spacers: say	£20.00
	£583.50
Labour	
Fixing: qualified steel fixer 30 hours × £7.10	£213.00
	£796.50
Profit and oncost 20%	£159.30
Rate per tonne	£955.80

Example 6.7

20 mm diameter hot rolled deformed high yield steel bars in isolated beams, straight or bent.	tonne

Materials	
1000 kg of 25 mm bars in stock lengths	£360.00
Waste: add 1%	£3.60
	£363.60
Unload and stack: labourer 1 hour × £6.50	£6.50
	£370.10
Tie wire: 7 kg × £17.80	£124.60
Chairs and spacers: say	£26.00
	£520.70
Labour	
Cutting, bending and fixing: qualified steel fixer 55 hours × £7.10	£390.50
	£911.20
Profit and oncost 20%	£182.24
Rate per tonne	£1093.44

Fabric reinforcement

Table 6.6 gives labour constants for unloading, stacking, cutting and laying fabric reinforcement.

Table 6.6 *Labour constants for fixing fabric reinforcement (labourer hours per unit)*

Concrete member or operation	Fabric reinforcement (kg/m^2)										
	0.77	1.54	2.61	3.02	3.73	4.34	4.53	5.55	6.16	6.72	8.14
Unloading, stacking, square cutting, laying (per m^2)											
Foundations	–	–	0.07	0.08	0.09	0.10	0.10	0.11	0.12	–	–
Beds and slabs	–	–	0.06	0.06	0.07	0.09	0.10	0.11	0.12	0.13	0.14
Walls	–	–	0.12	0.14	0.16	0.19	0.22	0.26	0.30	0.34	0.40
Staircases	–	–	0.16	0.20	0.25	0.31	0.37	0.44	0.50	0.56	0.65
Beam wrapping	0.15	0.19	–	–	–	–	–	–	–	–	–
Additional cutting (per metre)											
Raking cutting	0.05	0.06	0.08	0.09	0.11	0.13	0.14	0.15	0.16	0.17	0.18
Curved cutting	0.08	0.08	0.12	0.14	0.17	0.20	0.21	0.23	0.24	0.25	0.27

Example 6.8

Reinforcement; fabric; hard plain round steel, welded; reference A193; 200 mm side laps; 200 mm end laps	m^2

Materials

A193 fabric reinforcement: 1 m^2 at £3.02 kg/m^2 at £433.00/t	£1.31
Allowance for laps: add 12.5%	£0.16
Waste: add 5%	£0.07
	£1.54
Tie wire: 0.01 kg × £17.80	£0.18
Chairs and spacers: say	£0.95
	£2.67

Labour

Cutting and fixing: labourer 0.08 hours × £6.50	£0.52
	£3.19
Profit and oncost 20%	£0.64
Rate per m^2	£3.83

Example 6.9

Raking cutting on A193 fabric reinforcement.	m

Materials
Fabric reinforcement wasted by raking cutting: 0.25 m^2 at £1.31 £0.33

Labour
Cutting: labourer 0.09 hours × £6.50 £0.59

	£0.92
Profit and oncost 20%	£0.18
Rate per m	£1.10

Formwork

The materials employed to construct formwork for *in situ* concrete are generally softwood boards, plywood and sheet steel for working faces, supported on a softwood or steel framework, often in conjunction with proprietary props. Timber is still widely used in the making of formwork because of the variety of forms that it will allow. Plywood has generally superseded timber boards for forming the faces of members.

Sheets or boards must be of sufficient thickness to take the weight of wet concrete. For plywood the thickness ranges from 25 to 50 mm depending on the structures to be formed and the spans involved.

Timber forms for *in situ* work are usually unfit for further use after four to six uses. In the production of precast units in a factory, timber formwork will have up to twenty uses.

In accordance with the Working Rules, a carpenter required to reuse materials for concrete work is entitled to an extra payment.

Examples 6.10–6.12 refer to the formwork structure shown in Figure 6.2.

Example 6.10

Formwork to soffit of horizontal slabs; not exceeding 200 mm thick; height to soffit exceeding 3 m but not exceeding 4.5 m.	m^2

Area of soffit formwork between beams
$= (4000 - (2 \times 200)) \times (5000 - (2 \times 200)) = 16.56\,m^2$

Materials

Plywood 19 mm: 16.56 m^2 at £12.30	£203.69
Softwood:	
175 × 50 mm joists: 22 m at £1.88	£41.36
100 × 25 mm bearers: 9 m at £0.65	£5.85

The props are taken in the beams only (Example 6.11), but their costs could be proportioned between floors and beams if felt necessary.

Nails: 0.5 kg at £4.89	£2.45
	£253.35
Waste: add 10%	£25.34
	£278.69
Estimate 6 uses: cost per use 278.69/6	£46.45
Shutter oil to formwork: say	£2.20
Repairs and cleaning formwork: say	£2.10
Cost per 16.56 m^2	£50.75
Material cost per m^2: £50.75/16.56	£3.06
Labour	
Making, fixing and stripping: carpenter 2.7 hours at £8.80	£23.76
	£26.82
Profit and oncost 20%	£5.36
Rate per m^2	£32.18

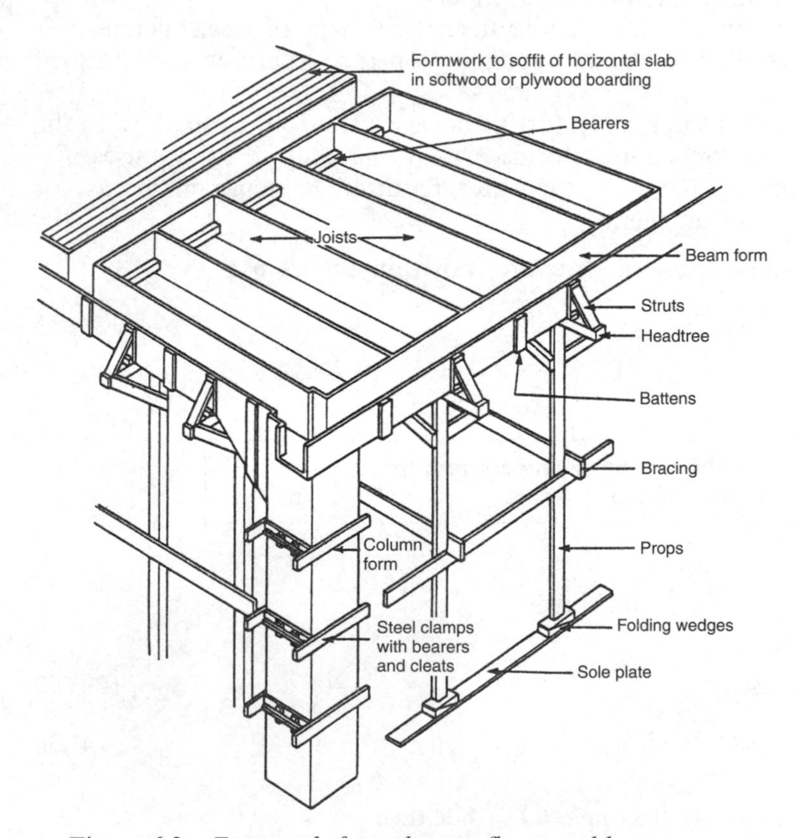

Figure 6.2 *Formwork for columns, floors and beams*

Example 6.11

Formwork to beams; attached to slabs; 400 × 400 mm; height to soffit exceeding 3 m but not exceeding 4.5 m.	m^2

Area of beam formwork = 3 × 4600 × 400 − 5.52 m^2

Materials	
Plywood 19 mm: 5.52 m^2 at £12.30	£67.90
Softwood:	
225 × 47 sole plate: 9 m at £2.98	£26.82
100 × 25 mm bearers: 6 m at £0.65	£3.90
100 × 100 mm props: 7 m at £3.10	£21.70
100 × 100 mm headtree 3 m at £3.10	£9.30
100 × 38 mm struts and battens: 5 m at £1.14	£5.70
Folding wedges: 1 m at £1.14	£1.14
Nails: 1 kg at £4.89	£4.89
	£141.35
Waste: add 10%	£14.14
	£155.49
Estimate 5 uses: cost per use 155.49/5	£31.10
Shutter oil to formwork: say	£1.40
Repairs and cleaning formwork: say	£1.70
Cost per 16.56 m^2	£34.20
Material cost per m^2: £34.20/5.52	£6.20
Labour	
Making, fixing and stripping: carpenter 3.9 hours at £8.80	£34.32
	£40.52
Profit and oncost 20%	£8.10
Rate per m^2	£48.62

Example 6.12

Formwork to isolated columns; 400 × 400 mm; height to soffit exceeding 3 m but not exceeding 4.5 m.	m^2

Area of beam formwork = 4 × 3500 × 400 = 5.6 m^2

Materials	
Plywood 19 mm: 5.6 m^2 at £12.30	£68.88
Softwood:	
100 × 25 mm cleats: 6 m at £0.65	£3.90
50 × 25 mm bearers: 7 m at £1.02	£7.14

Nails: 1 kg at £4.89	£4.89
	£84.81
Waste: add 10%	£8.48
	£93.29
Estimate 8 uses: cost per use £93.29/8	£11.66
Steel clamps: 7 set at £12.40 = £86.80	
Estimate 40 uses: cost per use £86.80/40	£2.17
Shutter oil to formwork: say	£0.90
Repairs and cleaning formwork: say	£1.70
Cost per 5.6 m^2	£16.43
Material cost per m^2: £16.43/5.6	£2.93
Labour	
Making, fixing and stripping: carpenter 4.2 hours at £8.80	£36.96
	£39.89
Profit and oncost 20%	£7.98
Rate per m^2	£47.87

Precast concrete

As stated in SMM7 in the coverage rules for precast concrete members, unit rates are deemed to include moulds, reinforcement, bedding, fixings, and so on. The unit rates developed in the examples that follow are for regular sections; however, the methods can be adapted to develop costs for irregular or one-off units.

The costs are for members produced on site by the contractor; therefore the rate build-ups used previously are adopted again.

Example 6.13

Before calculating the rates for individual precast concrete lintels, it is normal practice to calculate the cost of 1 m^3 concrete for various types of lintels itemized in the bills of quantities; for example, plain rectangular lintels, all faces left rough, reinforced with two 10 mm hot rolled plain mild steel bars. One hundred lintels, 100 × 100 mm × 1 m, can be cast from 1 m^3 concrete.

Concrete
The following costs per m^3 are taken from Example 6.4:

Materials	£72.66
Waste: add 5%	£3.63
Mixer	£0.41
Mixing squad	£3.48

Labour placing concrete in mould: 10 hours at £6.50		£65.00
Steel bars 10 mm: 200 m at 0.616 kg/m at £0.35/kg		£43.12
		£188.30
Mould		
Plywood 19 mm: 0.32 m^2 at £12.30	£3.94	
Mould clamps: 3 m at £0.40	£1.20	
100 × 25 mm cleats: 6 m at £0.65	£3.90	
50 × 25 mm bearers: 7 m at £1.02	£7.14	
Nails: 0.1 kg at £4.35	£0.44	
	£16.61	
Waste: add 2.5%	£0.42	
Carpenter making mould: 7 hours at £8.00	£56.00	
	£73.03	
Allow 20 uses per mould: cost per use £73.03/20 = £3.65		
At 1 use per lintel for 100 lintels: 100 × £3.65		£365.13
Cost to cost 100 Nr lintels 1 m long (1 m^3)		£553.43

Having established the cost of 1 m^3 of material for plain rectangular section lintels, we can calculate the cost for lintels of a similar pattern but with different size of section by using the ratio of area of cross-section to 1 m^2. This is illustrated in the next example.

Example 6.14

Lintels: plain rectangular shape; reinforced concrete; 100 × 150 × 1000 mm long; reinforced with 1.23 kg, 10 mm bars.	Nr

Materials	
Cost of 1 m^3 lintels 1 m long is (as Example 6.13) £553.43	
Cost of a 100 × 150 mm lintel 1 m long is 553.43 × (100 × 150)/(1000 × 1000)	£8.30
Mortar bed for lintels (inc. waste): say	£0.38
Labour	
Labourer stacking on site: 0.08 hours at £6.50	£0.52
Labourer transferring lintel from stack and hoisting to position on scaffold: 0.10 hours at £6.50	£0.65
Bricklayer and labourer lifting into position, levelling and bedding: 0.16 hours at £14.50	£2.32
	£12.17
Profit and oncost 20%	£2.43
Rate per m	£14.60
Rate per lintel 1 m long: £14.60 × (1000/1000)	£14.60

Example 6.15

Sills; sunk weathered; grooved; finished smooth 190 mm girth; 175 × 75 × 600 mm long; reinforced 0.37 kg, 10 mm bars.	Nr

In this example we shall presume that the sills have been purchased from a manufacturer and delivered to the site.

Materials	
Cost of 175 × 75 × 600 mm sill delivered to site	£18.90
Mortar bed for lintels (inc. waste): say	£0.28
Labour	
Labourer unloading and stacking on site: 0.07 hours at £6.50	£0.46
Labourer transferring lintel from stack and hoisting to position on scaffold: 0.10 hours at £6.50	£0.65
Bricklayer and labourer lifting into position, levelling and bedding: 0.13 hours at £9.88	£1.28
	£21.57
Profit and oncost 20%	£4.31
Rate per sill	£25.88

7

Brickwork and blockwork

SMM7 Sections F10, F30

Introduction

Bricks can be obtained in a number of sizes and various compositions, strengths and shapes, to mention only the more obvious factors. The appropriate British Standards give details for clay, concrete, sand-lime and other bricks.

To give examples of rates using all of the different types of bricks would be impossible in the space allocated, as well as highly repetitive. Therefore the examples in this chapter will concentrate on a metric brick with dimensions 215 mm long, 102.5 mm wide and 65 mm thick. Whether the brick is solid, hollow or perforated, or has single or double frogs, will make no difference to us here, although in theory the different weights of bricks could affect the laying output per man and other factors.

Mortar

If bricks have no perforations or frogs, are absolutely rectilinear, are laid on a bed of mortar 10 mm thick and have each end jointed 10 mm thick, then 1000 bricks require the following volume of mortar:

$$1000[(0.215 \times 0.1025 \times 0.01) + (0.1025 \times 0.065 \times 0.01)] = 0.287\,\text{m}^3$$

The amount of mortar taken by estimators varies from 0.5 to 0.8 m^3. This makes allowance for the mortar which disappears into the frogs, hollows or perforations in the bricks, and also for the large amount of waste generated. This waste occurs throughout the process: cement, lime and sand are dumped on the ground and trodden in; mortar is left in the drum of the mixer, in the barrow, on the mortar board and on the scaffold; and mortar is squeezed out of the bed and joint of every brick. The allowance is large, but there is a lot of waste. The mixes and costs of mortars used in the examples in this chapter are given in Appendix A at the end of the book.

Bricks per wall area

Obviously we need to know the number of bricks to be laid in each square metre of wall. For a half brick wall laid using metric bricks with dimensions 215 × 102.5 × 65 mm and mortar joints of 10 mm, the number of bricks required per square metre is given by:

$$\frac{1000 \times 1000}{225 \times 75} = 59.25925, \text{ say } 59$$

One brick walls need 118 bricks; one and a half brick walls need 177 bricks; and so on. The waste on bricks can vary as with any other material, but in this chapter we will allow a modest 5 per cent.

Labour costs

Labour costs depend first on the squad make-up, i.e. the numbers of craftsmen and labourers. In the examples in this chapter, the squad comprises two bricklayers and one labourer.

The second factor is the number of bricks which each bricklayer can lay per hour. Fanciful claims are common on building sites; tales abound of bonuses earned, records shattered and employers dumbfounded. The truth is much less in magnitude. A good average – every day, 5 days a week, 47 weeks per annum – is 55–65 common bricks per hour. Obviously this is in straight runs of walling without excessive cutting for bond and without anything unusual. Output might be a little higher on thicker walls simply because the bricklayers and their labourer are spending proportionately less time moving along the wall.

Common brickwork

Example 7.1

Common brickwork half brick thick in cement mortar 1:3, built in stretcher bond.	m^2

Common bricks 215 × 102.5 × 65 mm delivered to site: per 1000		£125.00
Wastage 5%		£6.25
		£131.25
Labour: 2 bricklayers per hour at £8.00	£16.00	
1 labourer per hour at £6.50	£6.50	
Squad cost	£22.50	
Output per tradesman 55 bricks/hour; 2 tradesmen:		
Cost labour per 1000 bricks: £22.50 × 1000/(2 × 55)		£204.55
Mortar at £109/m^3 at 0.53 m^3 per 1000 bricks for half brick wall		£57.77
Cost for 1000 bricks		£393.57
Number bricks per m^2 for half brick thick wall is 59		
Cost per m^2: £393.57 ×(59/1000)		£23.22
Profit and oncost 20%		£4.64
Rate per m^2		£27.86

Example 7.2

Common brickwork one brick thick in cement mortar 1:3, built in English bond.	m^2

Common bricks 215 × 102.5 × 65 mm delivered to site: per 1000		£125.00
Wastage 5%		£6.25
		£131.25
Labour: 2 bricklayers per hour at £8.00	£16.00	
1 labourer per hour at £6.50	£6.50	
Squad cost	£22.50	
Output per tradesman 55 bricks/hour; 2 tradesmen:		
Cost labour per 1000 bricks: £22.50 × 1000/(2 × 55)		£204.55
Mortar at £109/m^3 at 0.53 m^3 per 1000 bricks for half brick wall		£57.77
Cost for 1000 bricks		£393.57
Number bricks per m^2 for one brick thick wall is 118		
Cost per m^2: £393.57 × (118/1000)		£46.44
Profit and oncost 20%		£9.29
Rate per m^2		£55.73

Example 7.3

Common brickwork one and one half brick thick in cement/lime mortar 1:6, built in Flemish bond.	m^2

Common bricks 215 × 102.5 × 65 mm delivered to site: per 1000		£125.00
Wastage 5%		£6.25
		£131.25
Labour: 2 bricklayers per hour at £8.00	£16.00	
1 labourer per hour at £6.50	£6.50	
Squad cost	£22.50	
Output per tradesman 60 bricks/hour; 2 tradesmen:		
Cost labour per 1000 bricks:		£187.50
Mortar at £99/m^3 at 0.60 m^3 per 1000 bricks for half brick wall		£59.40
Cost for 1000 bricks		£378.15
Number bricks per m^2 for 1.5 brick thick wall is 177		
Cost per m^2: £378.15 × (177/1000)		£66.93
Profit and oncost 20%		£13.39
Rate per m^2		£80.32

Brickwork sundries

Example 7.4

Form cavity 75 mm wide including four butterfly wall ties per m^2.	m^2

Cost 4 wall ties per m^2 at £46.50 per 1000	£0.19
Wastage 5%	£0.01
Tradesman sets out cavity at 12 m^2/hour at £8.00/hour:per m^2	£0.67
	£0.86
Profit and oncost 20%	£0.17
Rate per m^2	£1.03

Example 7.5

> Hessian based bituminous sheet complying with BS 743 Type A in horizontal damp proof course, lapped at joints and corners, in width not exceeding 225 mm. m^2

Most authorities apply a standard cost for laying this material irrespective of width, although SMM7 does require the quantity surveyor to keep the classifications separate. The labour output for vertical damp proof course (DPC) is half that for horizontal DPC.

DPC: per m^2	£4.45
Waste and laps 5%	£0.22
Tradesman at £8.00 per hour lays 5 m^2:per m^2	£1.60
Labourer at £6.50 per hour lays 5 m^2:per m^2	£1.30
	£7.57
Profit and oncost 20%	£1.51
Rate per m^2	£9.09
Rate per m at 225 mm wide	£2.04

Example 7.6

> Hessian based bituminous sheet in complying with BS 743 Type A vertical damp proof course, lapped at joints and corners, in width not exceeding 225 mm. m^2

DPC: per m^2	£4.45
Waste and laps 5%	£0.22
Tradesman at £8.00 per hour lays 2.5 m^2:per m^2	£3.20
Labourer at £6.50 per hour lays 2.5 m^2:per m^2	£2.60
	£10.47
Profit and oncost 20%	£2.09
Rate per m^2	£12.57
Rate per m at 225 mm wide	£2.83

Example 7.7

Galvanized expanded metal brickwork reinforcement, 100 mm wide, bedded in mortar.	m^2

Material per 25 m roll 100 mm wide (1 roll = 2.5 m^2)	£15.60
Waste 5%	£0.78
Tradesman at £8.00 per hour lays 2.5 m^2:per m^2	£3.20
Labourer at £6.50 per hour lays 2.5 m^2:per m^2	£2.60
	£22.18
Profit and oncost 20%	£4.44
Rate per roll	£26.62
Rate per m at 100 mm wide	£1.06

Facing brickwork

Example 7.8

One and one half brick thick wall built with common bricks, faced one side with facing bricks PC £380 per 1000 delivered site, built in Flemish bond and jointed with in cement/lime mortar 1:1:6.	m^2

Cost of facing bricks per 1000 delivered to site		£380.00
Unload and stack: labourer 2 hours at £6.50		£13.00
		£393.00
Wastage 5%		£19.65
Labour: 2 bricklayers per hour at £8.00	£16.00	
1 labourer per hour at £6.50	£6.50	
Squad cost	£22.50	
Output per tradesman 40 bricks/hour: 2 tradesmen		
Cost labour per 1000 bricks: squad cost × 1000/(2 × 40)		£281.25
Mortar at £99/m^3 at 0.60 m^3 per 1000 bricks for 1.5 brick thick wall		£59.40
Total cost for building 1000 facing bricks		£753.30
Number of facing bricks per m^2 in Flemish bond is 79		
Cost of facing brick per m^2: cost/thou × (79/1000)		£59.51
Bricklayer joints 2.4 m^2/hour at £8.00: per m^2		£3.33
Number of bricks per m^2 for 1.5 brick wall is 177: so commons 177 − 79 = 98		
Commons in same mortar cost £378.15 per 1000 (Example 7.3): 98 cost		£37.06
		£99.90

Profit and oncost 20%	£19.98
Rate per m^2	£119.88

Example 7.9

> Plain brick-on-edge horizontal cope one and one half brick thick wide, formed with facing bricks PC £380 per 1000 delivered site, built and jointed in cement / lime mortar 1:1:6. m

Cost of facing bricks per 1000 delivered to site		£380.00
Unload and stack: labourer 2 hours at £6.50		£13.00
		£393.00
Wastage 5%		£19.65
Labour: 2 bricklayers per hour at £8.00	£16.00	
1 labourer per hour at £6.50	£6.50	
Squad cost	£22.50	
Output per tradesman 40 bricks / hour: 2 tradesmen		
Cost labour per 1000 bricks: squad cost × 1000/(2 × 40)		£281.25
Mortar at £99/m^3 at 0.60 m^3 per 1000 bricks for 1.5 brick thick wall		£59.40
Total cost for building 1000 facing bricks		£753.30
Cope with 65 mm thick bricks: 1 m /(65 + 10) = say 14 bricks		
Number of bricks per metre $1\frac{1}{2}$ brick wide is 21		
Cost of facing brick per m		£35.87
Bricklayer joints 2.4 m^2 / hour at £8.00: per m^2		
Girth of cope 112.5 + 327.5 + 112.5 = 553 mm		
Area jointed per metre is 0.55 m^2: cost £8.00 × (0.55/2.4)		£1.83
		£37.70
Profit and oncost 20%		£7.54
Rate per metre		£45.25

Jointing and pointing

In Examples 7.8 and 7.9 the brickwork was described as jointed. This is the term applied to the finishing of the joints between bricks, usually with a struck weathered joint using the same mortar as for building. The most important feature of jointing is that it is done as the bricks are built, the mortar in each bed or joint being formed while still plastic into the struck weathered finish. No material is added at a later date.

Pointing is the term used to describe the procedure where a separate mortar – very often of a different mix and even colour from the building mortar – is applied to the joints between the bricks. As the bricklayer lays the bricks he will rake out a little of the bedding mortar to leave space for this additional application. The finish achieved at the joints may be struck weathered but there are many other finishes, and the reader is referred to the many books on brickwork currently available.

In Example 7.8 we could add pointing with a white mortar made from Snowcrete cement 1:4. We have no calculation for this mortar but will price arbitrarily at £140.00 per m^3. The calculation for the new rate would therefore be as follows.

Example 7.10

One and one half brick thick wall built with common bricks, faced one side with facing bricks PC £380 per 1000 delivered site, built in Flemish bond in cement/lime mortar 1:1:6 and pointed in white mortar made from Snowcrete and silver sand 1:4.	m^2

Cost of bricks as Example 7.8	£412.65
Cost of labour as Example 7.8	£281.25
Mortar to build as Example 7.8	£59.40
Mortar to point 1000 bricks: 0.033 m^3 at £140 per m^3	£4.62
Total cost of building 1000 facing bricks	£757.92
Number of facing bricks per m^2 in Flemish bond is 79	
Cost of facing brick per m^2: cost per thou × (79/1000)	£59.88
Bricklayer rakes and points joints 1 m^2 hour £8.00	£8.00
	£67.88
Cost 98 commons as Example 7.8	£37.06
	£104.93
Profit and oncost 20%	£14.53
Rate per m^2	£119.46

Special brickwork

Example 7.11

Facing brickwork in horizontal projection 56 mm deep from the face of the wall and 440 mm high, built brick-on-end in cement/lime mortar 1:1:6.	m^2

Cost of facing bricks per 1000 including waste as Example 7.8	£412.65
Squad cost £22.50 per hour as Example 7.8	
Cost labour per 1000 bricks: squad cost × 1000/(2 × 25)	£450.00
Mortar at £99.00/m^3 at 0.60 m^3 per 1000 bricks for 1.5 brick wall	£59.40
Total cost of building 1000 facing bricks	£922.05

Projection with 65 mm thick bricks has 26.66 bricks/m, say 27 Cost is cost of 100 bricks × (27/1000)	£24.90

To produce a projection of 56 mm requires that a cut brick be built into the wall behind the projecting facing brick. The cut bricks are of course common bricks. It is assumed that one common brick will not be cut into two usable pieces; this covers waste in the cut brickwork.

Commons in same mortar cost per 1000 as Example 7.3:27 cost	£10.21
Additional labour rough cutting common brickwork 0.44 m^2 per m of projection	
Bricklayer cuts 1 m^2/ hour at £8.00/hour	£3.52
Bricklayer joints 2.4 m^2/ hour at £8.00 Girth of projection $440 + (2 \times 56) = 552$ mm Area jointed per metre is 0.55 m^2: cost of tradesman × (0.55/2.4)	£1.83
	£40.46
Profit and oncost 20%	£8.09
Rate per metre	£16.18

> Note to Example 7.11: To produce a projection of 56 mm requires that a cut brick be built into the wall behind the projecting facing brick. The cut bricks are of course common bricks. It is assumed that one common brick will not be cut into two usable pieces; this covers waste in the cut brickwork.

Example 7.12

> Facing brick semi-circular arch, one brick wide on face and one and one half bricks wide on soffit, with purpose made facing brick voussoirs, £650.00 per 1000 delivered site, built and jointed in cement / lime mortar 1:1:6. m^2

Cost of facing bricks per 1000 delivered to site	£650.00
Unload and stack: labourer 2 hours at £6.50	£13.00
	£663.00
Wastage 5%	£33.15
	£696.15
Squad cost £22.50 per hour as Example 7.8 Output per tradesman 20 bricks/ hour: 2 tradesmen Cost labour per 1000 bricks: squad cost × 1000/(2 × 20)	£562.50
Mortar at £99.00/m^3 at 0.60 m^3 per 1000 bricks for 1.5 brick wall	£59.40
Cost of laying 1000 voussoirs	£1318.05

Mean girth or length on face: assume that the bricks are 75 mm at the widest part (the extrados) and 55 mm wide at the narrowest part (the intrados). The mean width of each voussoir is then 65 mm. Allowing a 10 mm joint, the number of bricks per metre of arch is 13.33 for a half brick thick wall. Say 40 bricks per metre are required for a 1.5 brick arch.

Cost per metre is cost/thou × (40/1000)	£52.72
Bricklayer joints 1.5 m^2 per hour at £8.00 Girth of arch 327.5 + (2 × 225) = 777.5 mm (2 × faces and soffit) Area jointed per metre is 0.78 m^2: cost tradesman's rate × (0.78/1.5)	£4.16
	£56.88
Profit and oncost 20%	£11.38
Rate per metre	£68.26

Such an arch requires to be provided with support or 'centering' as it is built. Centering is priced in a similar fashion to formwork including 're-uses' where appropriate. Shuttering oil is not used as it would stain the masonry but there is often a layer of building paper or polythene placed over the centering which allows easy release from the mortar of the masonry and also prevents wet mortar or laitence dripping onto finished masonry below.

Blockwork

Blocks are manufactured in a variety of materials, clay and concrete being the most usual. They can be hollow, joggle jointed, and filled with foamed plastic, and may be finished on one or both faces with a variety of decorative finishes or left plain to receive plaster. Like bricks they can be built fair faced, and are very often painted direct when built this way. Building blocks with a fair face or a pre-applied decorative finish will lower output and will also introduce the need for jointing or pointing. The latter are treated in the same way as for brickwork; the bricklayer can point a larger area of blockwork per hour than of brickwork.

Example 7.13

75 mm thick lightweight aggregate concrete block complying with BS 6073: Part 1, built in cement/lime mortar 1:1:6.	m^2

Material per m^2 delivered to site and MOL	£6.12
Waste 5%	£0.24
Squad cost £22.50 per hour as Example 7.8 Output per bricklayer 1.5 m^2/hour: 2 bricklayers Cost labour per m^2: squad cost/(2 × 1.5)	£7.50
Mortar at £96.00/m^3 at 0.007 m^3 per m^2 of 75 mm blockwork	£0.67
	£14.53
Profit and oncost 20%	£2.91
Rate per m^2	£17.44

Example 7.14

100 mm thick autoclaved aerated concrete block complying with BS 6073: Part 1, built in cement/lime mortar 1:2:9.	m^2

Material per m^2 delivered to site and MOL	£7.36
Waste 5%	£0.37
Squad cost £22.50 per hour as Example 7.8	
Output per bricklayer 1.5 m^2/hour: 2 bricklayers	
Cost labour per m^2: squad cost/(2 × 1.5)	£7.50
Mortar at £96.00/m^3 at 0.01 m^3 per m^2 of 100 mm blockwork	£0.96
	£16.19
Profit and oncost 20%	£3.24
Rate per m^2	£19.43

8

inning

ıs C30, D50

In

Un a building to a lower stratum of the ground.
Thi

- T nt within or adjacent to the existing foun-
  d basement.
- T l than the existing, where the existing has
  fa ubsidence.
- T xisting foundation.
- Ir to be mechanically moved by installing a
  g

Ther derpinning operation; the choice depends
upor ɔt in the most simple cases and where the
struc arried out by a specialist contractor. This
cont as the drilling of ground anchors and/or
the c

Fo ucture the contractor will price the work
descr s used will be based upon those for the
grou ıs.

Shor

It may be nece to price an item in the bills of quantities for strutting to window and door openings as well as for temporary shoring to walls. The extent of this work will be described as an item and it will be necessary for the estimator to have this temporary structural work designed before pricing.

Underpinning (D50)

The excavation work in underpinning is in two parts: the preliminary trench from the existing surface level to the base of the foundation, and the underpinning pits below the existing foundation. The preliminary trench will be priced in an identical manner to that described for hand excavation in Chapter 5.

The underpinning pits are the excavations below the existing foundation level undertaken by hand on a section of foundation normally 1 m to 1.5 m long. The working on such small sections, in part below the existing foundation, results in an addition to the hand excavation constant of 100 per cent. The bringing of the soil to the surface in the number of lifts calculated by dividing the total depth to the base of the underpinning by 1.5 m, as in Chapter 5, is calculated at the given constants plus 50 per cent.

Where underpinning is carried out by working from one side only, the earthwork support to the back face of the excavation under the existing foundation is left in. This section of earthwork support is commonly in precast concrete planks, each with a hole in the top to allow for grouting any void left behind the earthwork support (see Example 8.2).

For all remaining items in the underpinning section, i.e. groundwork, *in situ* concrete and masonry, the constants used are those of the appropriate chapter plus 50 per cent.

Example 8.1

Excavating underpinning pits, maximum depth not exceeding 2 m, from one side only.	m^3

The difficulty multipliers are obtained from Table 5.2, and the outputs (man hours per m^3) from Tables 5.3 and 5.4.

	Volume excavated	Difficulty multiplier	Man hours	Rate per hour	Cost
Max. throw	1.5	3	2.2	£6.50	£64.35
Max. throw	0.5	3	2.2	£6.50	£21.45
Throw one	0.5	2	1.4	£6.50	£9.10
Total volume	2			Total cost	£94.90
Dividing total cost by total volume = cost/m^3					£47.45
Profit and oncost 20%					£9.49
Rate per m^3					£56.94

Example 8.2

Earthwork support to underpinning pits left in, maximum depth not exceeding 2 m, distance between opposing faces 2–4 m.	m^2

Materials

Precast reinforced concrete plank 50 mm × 200 mm wide × 1 m long:				
5 nr at £8.00				£40.00
Waste 5%				£2.00
Walings 50 × 200mm:				
3 m × 1 × 0.05 × 0.20 m at 0.030		£280.00/m^3	£8.40	
Strutting 100 × 100 mm (to ground) at 2 per metre = 3 pairs:				
6 nr × (av.) 1.2 × 0.10 × 0.10 m at 0.072		£280.00/m^3	£20.16	
Base 100 × 100 × 200 mm long (base of struts):				
6 × 0.02 × 0.10 × 0.30 m at 0.012		£280.00/m^3	£3.36	
	0.114		£31.92	

Divided by 3 uses			£10.64	
Waste 10%			£1.06	£11.70
Labour				
Unload, store and place in position precast concrete planks:				
5 nr at 12 minutes each at	1.00	£6.50/hour		£6.50
Unload, store, set up and strike timberwork:				
10 hours/m^3	1.14	£6.80/hour		£7.75
				£67.96
Profit and oncost 20%				£13.59
Rate per m^2				£81.55

9

Roof coverings

SMM7 Sections H60, H62, H63, H71

Slate roofing

The main sources for roofing slates in Britain are Cumbria, Cornwall and North Lancashire in England, and Bangor, Portmadoc and Caernarvon in Wales. There are no longer large scale quarries in Scotland; therefore the main source of Scottish slates is from demolition for use second hand. On the British market there are also roofing slates from European countries such as Spain and Portugal. Roofing slates are sold by the tonne or by the thousand depending upon the source.

British Standard BS680 Part 1:1944 (imperial) and Part 2:1971 (metric) details geological formations of true slate rock from which roofing slates can be quarried, together with characteristics, grade tests and sizes. The labour outputs given here are for standard size slates in accordance with BS 680 Part 2 Table 1 'Standard lengths and widths of slates'. Sizes of randoms and peggies can be obtained from Table 2 'Range of lengths for randoms and peggies'.

The labour outputs in Table 9.1 are based on a squad of two slaters and one labourer, and allow for carrying and fixing in position. Unloading, holing and dressing of slates are highlighted as required in the examples, which follow.

Table 9.1 *Range of slater outputs fixing slates*

Slate size (mm)	Output per tradesman (m^2/hour)	Slate size (mm)	Output per tradesman (m^2/hour)
610 × 355		355 × 205	
610 × 305	5.0–6.0	355 × 180	1.5–2.5
560 × 305		330 × 280	
560 × 280		330 × 255	
510 × 305	4.0–4.5	330 × 205	1.5–2.5
510 × 255		330 × 180	
460 × 305		305 × 255	
460 × 255	2.0–3.0	305 × 205	1.5–2.0
460 × 230		305 × 155	
405 × 305		255 × 255	
405 × 255	2.0–3.0	255 × 205	1.25–1.5
405 × 230		255 × 150	
405 × 205			
355 × 305	2.0–3.0		
355 × 255			

The wastage on slates is taken as 2.5 per cent for sized and 5 per cent for unsized slates. Under the rules laid down in SMM7, roof and wall coverings are measured in square metres, and items are deemed to include underlay and battens. These items have therefore to be accounted for under the superficial items for slating and tiling where required.

Tables 9.2 and 9.3 show expected slater outputs for a range of tasks.

In the following examples, in order to calculate the cost per m^2 we have first to establish the covering capacity of one slate. To do this we have to calculate the gauge (fixing centres) of the slate and multiply by its width.

$$\text{For head nailed slates, the gauge} = \frac{\text{average length} - (\text{lap} + 25\,\text{mm})}{2}$$

$$\text{For centre nailed slates, the gauge} = \frac{\text{average length} - \text{lap}}{2}$$

Gauge may also be defined as the length of slate visible.

Table 9.2 *Range of slater outputs for labour items*

Measured items	Output per hour per tradesman
Square cutting around openings	3–4 m
Angle cutting to valleys	4–6 m
Angle cutting to hips	8–12 m
Double course at eaves	14–18 m
Square verges including bedding	6–10 m
Holes for pipes	3–4 Nr
Ridge capping including bedding	5–6 m
Hip capping including bedding	4–5 m
Wastage on labour items:	
Square cutting	Add 33.33%
Raking cutting	Add 50%

Table 9.3 *Range of slater outputs for associated works*

Associated works	Output per hour per tradesman
Underslating felt	2.5–3.5 rolls of 15 m^2
Battens	35–45 m

Example 9.1

Roofing; best quality second hand Scottish slates to BS 680; random sizes 355 to 255 mm long in diminishing courses; fixing with two 40 mm galvanized steel nails per slate to 75 mm lap on reinforced underslating felt to softwood sawn sarking, pressure impregnated with preservative; coverings to sloping pitch 40 degrees.	m^2

Materials: slates

Second hand slate delivered to site at £350.00 per 1000	£350.00
Take delivery and stack: 1.5 hours per 1000 at £6.50	£9.75
Nails: 40 mm galvanized at £4.46/kg at 3.18 kg per 1000: 2000 nails per 1000 slates	£28.37
	£388.12
Allow waste 5%	£19.41
Labour preparing slates per 1000:	
Tradesman reholing: 4 hours; redressing: 2 hours; 6 hours at £8.00	£48.00
Labourer sizing: 2 hours at £6.50	£13.00
Cost per 1000 slates prepared ready for fixing	£468.53

Materials: fixing slates

Slate gauge, head nailed = [average length − (lap+25)]/2 mm

= [305 − (75 + 25)]/2 = 102.5 mm = 0.1025 m

Coverage 1 slate = width × gauge = 0.205 × 0.1025 = 0.021 m^2

Coverage 1000 slates = 0.021 × 1000 = 21.01 m^2

Cost per m^2 = £468.53/21.01 — £22.30

Labour: fixing slates

From Table 9.1, for the average slate size of 305 × 205 mm, 1.75 m^2 can be carried and fixed in position per hour per tradesman. Assume a squad of 2 tradesman and 1 labourer.

Labour cost per hour: (2 × £8.00) + (1 × £6.50) = £22.50

Cost per m^2: 0.29 × £22.50 — £6.43

Slating cost per m^2 — £28.73

In accordance with SMM7 coverage rules for coverings, underlay and battens are deemed to be included. Therefore, where a specification requires either of those items, their costs have to be incorporated in the rate for the coverings as follows.

Materials: underslating felt

Felt 15 m^2 roll delivered to site	£22.47
Unload and stack: 0.05 hours at £6.50	£0.33
	£22.80

Nails fixed at 300 mm centres: roll length 15 m + width 1 m = 16 m

Number of nails per roll: 16/0.3=53.33 (say 54)	
Nails: 20 mm galvanised at £4.46/kg at 2.85 kg per 1000: 54 nails per roll	£0.69
	£23.49
Allow waste 5%	£1.17
	£24.66

Labour: fixing underslating felt

Squad of 2 tradesmen and 1 labourer fixes 6 rolls per hour	
Labour cost per hour: £22.50	
Cost to fix 1 roll	£3.75
Cost to lay 1 roll	£28.41

To calculate the cost for 1 m^2 of felt we must first determine the effective coverage of 1 roll. That is, we must subtract the laps from the roll dimensions, as follows:

Coverage $= (15 - 0.15) \times (1 - 0.075) = 14.85 \times 0.925 = 13.74\,m^2$	
Cost to lay 1 m^2 felt: £28.41/13.74	£2.07
	£30.80
Profit and oncost 20%	£6.16
Rate per m^2	£36.96

Example 9.2

> Roofing; natural slates, 'Penrhyn' heather blue standard grade; size 405 to 255 mm; fixing with two copper nails per slate to 75 mm lap to softwood battens 38 × 25 mm pressure impregnated with preservative, on reinforced underslating felt; coverings to sloping pitch 30 degrees. m^2

Materials: slates

Heather blue slate ex quarry at £780.00 per 1000		£780.00
Assume site 200 miles from quarry		
700 slates per crate weighing 1.06 tonnes per crate: by 16 tonne lorry		
Nr of crates in load: 16/1.06 = 15.09, say 15 crates		
Nr of slates in load: 15 × 700 = 10 500		
Loading lorry:	1	
Unloading at site:	1	
Travelling time 1 man:	10	
Rest time etc.:	8	
	20 hours at £6.80	£136.00
Lorry cost: 20 hours at £23.80		£476.00
Delivery cost of full load 10 500 slates		£612.00

Delivery cost for 1000 slates: £612.00/10.5	£58.29
Take delivery and stack: 1.5 hours per 1000 at £6.50	£9.75
Nails: 40 mm copper at £6.89/kg at 2.7 kg per 1000: 2000 nails per 1000 slates	£37.21
	£885.25
Allow waste 5%	£44.26

Materials: fixing slates
Labour preparing slates per 1000:

Tradesman holing: 4 hours at £8.00	£32.00
Cost per 1000 slates prepared ready for fixing	£961.51

Slate gauge, head nailed = [length − (lap + 25)]/2 mm
= [405 − (75 + 25)]/2 = 152.5 mm = 0.1525 m
Coverage 1 slate = width × gauge = 0.255 × 0.1525 = 0.039 m^2
Coverage 1000 slates = 0.039 × 1000 = 38.89 m^2
Cost per m^2 = £961.51/38.89 — £24.72

Labour: fixing slates
From Table 9.1, for the average slate size of 405 × 255 mm, 2.5 m^2 can be carried and fixed in position per hour per tradesman. Assume a squad of 2 tradesman and 1 labourer.
Labour cost per hour (2 × £8.00) + (1 × £6.50) = £22.50

Cost per m^2: 0.20 × £22.50	£4.50
Slating cost per m^2	£29.22

In accordance with SMM7 coverage rules for coverings, underlay and battens are deemed to be included. Therefore, where a specification requires either of those items, their costs have to be incorporated in the rate for the coverings as follows.

Underslating felt as Example 9.1: per m^2 — £2.07

Materials: battens
Nr of slates per m^2: 1000/38.89 = 25.72
Quantity 38 × 25 mm battens per m^2 = nr slates per m^2 × slate width
= 25.72 × 0.255 = 6.56 m

Cost of battens per m^2: 6.56 m at £0.46 per m	£3.02	
Nr nails per m^2 = (length battens)/(rafter centres) = 6.56/0.45		
Weight nails per m^2 at 260 nails/kg: (6.56/0.45)/260 = 0.06 kg		
Cost of nails per m^2: 0.06 kg at £3.56/kg	£0.21	
	£3.23	
Allow waste 5%	£0.16	£3.39

Labour: battens
Squad of 2 tradesmen and 1 labourer fixes 80 m battens per hour
Labour cost per hour: £22.50

Cost to fix battens for 1 m^2 slating: £22.50 × (6.56/80)	£1.85
	£36.53
Profit and oncost 20%	£7.31
Rate per m^2	£43.84

Example 9.3

Eaves; double course of 405 × 255 mm heather blue slates at eaves.	m

Materials: slates	
Cost 1000 heather blue slates from Example 9.2	£780.00
As the slates arrive on site pre-sized at 405 × 255 mm, those at eaves have to be dressed to a length equal to gauge + lap + 25 mm	
Dressing to size: 2 hours per 1000 at £8.00	£16.00
Cost for 1000 slates ready for fixing	£796.00
Materials: fixing slates and battens	
Nr slates of width 255 mm per 100 m eaves: 100/0.255 = 392.16	
Cost of slates per 100 m eaves: £796.00× (392.16/1000)	£312.16
Additional row of battens: 100 m at £0.46/m	£46.00
Nr nails = (length battens)/(rafter centres) = 100/0.45	
Weight of nails at 260 nails/kg: (100/0.45)/260 = 0.85 kg	
Cost of nails: 0.85 kg at £3.56/kg	£3.03
Labour: fixing slates and battens	
Nr of slates per m^2: 1000/38.89 = 25.72	
Area slates per 100 m of eaves: 392.16/25.72 = 15.25 m^2	
Labour cost per hour: (2 × £8.00) + (1 × £6.50) = £22.50	
Output per hour per tradesman: 2.5 m^2	
Cost to fix 100 m slates at eaves: £22.50 × 15.25/(2 × 2.5)	£68.61
Squad of 2 tradesmen and 1 labourer fixes 80 m of battens per hour	
Cost to fix 100 m battens: £22.50 × (100/80)	£28.13
	£457.93
Profit and oncost 20%	£91.59
Cost per 100 m	£549.52
Rate per metre	£5.50

Example 9.4

Verges; single extra undercloak course 115 mm wide; bedding and pointing in cement mortar 1:3.	m

Materials

Slate-and-a-half slates to alternate courses to form verge:		
Heather blue slate ex quarry at £1630.00 per 1000		£1630.00

Assume site 200 miles from quarry
500 slates per crate weighing 1.00 tonnes per crate: by 16 tonne lorry
Nr of crates in load: 16/1.00 = 16
Nr of slates in load: 16 × 500 = 8500

Loading lorry:	1.12		
Unloading at site:	1.12		
Travelling time 1 man:	10		
Rest time etc.:	8		
	20.24 hours at £6.80	£137.63	
Lorry cost: 20.24 hours at £23.80		£481.71	
Delivery cost of full load 10 500 slates		£619.34	
Delivery cost for 1000 slates: £619.34/8			£77.42
Take delivery and stack: 1.5 hours per 1000 at £6.50			£9.75
Nails: 40 mm copper at £6.89/kg at 2.7 kg per 1000: 2000 nails per 1000 slates			£37.21
			£1754.38
Allow waste 5%			£87.72

Labour preparing slates

Tradesman holing: 4 hours at £8.00	£32.00
Cost per 1000 slates prepared ready for fixing	£1874.10

Materials: fixing slates

Nr slates of gauge 153 mm per 100 m verge: 100/0.153 = 654		
Slate-and-a-half slate are used on alternate courses: 654/2 = 327		
Cost of slate-and-a-half slates: 327 × £1874.10 per 1000		£612.83
Cost of 405 × 255 mm slates: 327 × £961.51 per 1000		£314.41
Cost of slates per 100 m verge		£927.24
Deduct cost of slates measured in superficial item (Example 9.2)		
Coverage slate-and-a-half slates: 327 × (0.153 × 0.380)	19.01 m^2	
Coverage 405 × 255 mm slates: 327× (0.153 × 0.255)	12.76 m^2	
Area covered by verge	31.77 m^2	
Cost to deduct: 31.77 at £24.72 per m^2 (Example 9.2)		£785.35
Extra over cost of slates per 100 m verge		£141.89
Cement mortar 1:3: cost say £75.89 per m^3		
Mortar required to bed 100 m verge: 100 × 0.115 × 0.025 = 0.2875		
Cost of mortar: £75.89 × 0.2875		£21.82
Slate undercloak 115 mm wide: 100 m at £2.35/m		£235.00

Labour: fixing slates	
Forming square verges including bedding: 14.5 hours per 100 m at £8.00	£116.00
	£514.71
Profit and oncost 20%	£102.94
Cost per 100 m	£617.65
Rate per metre	£6.18

Example 9.5

Hips: raking; on 405 × 255 mm heather blue slates.	m

Using trigonometry we can establish the plane angle at the hip for a particular roof. From this we can calculate the length of the cut per slate, and thereafter the number of slates per metre which will be used in wastage calculations.

As an example, take a roof of 30 degrees pitch with a standard hipped end, spanning 8 m (Figure 9.1). First,

$$\tan 30^\circ = \frac{\text{height of rise of roof}}{\text{half span of roof}}$$

$$\text{height of rise} = 4 \times \tan 30^\circ = 2.31\,\text{m}$$

Next,

$$\text{rafter length} = \sqrt{(2.31^2 + 4^2)} = 4.62\,\text{m}$$

Denote the plane angle at the hip as α. Then

$$\tan\alpha = \frac{\text{rafter length}}{\text{half span of roof}} = \frac{4.62}{4} = 1.155$$

$$\alpha = 49.11^\circ$$

Now consider the cut slate:

$$\tan\alpha = \frac{\text{height of cut up longer edge of slate}}{\text{width of slate}}$$

$$\text{height of cut} = 1.155 \times 255 = 293\,\text{mm}$$

Then

$$\text{cut length} = \sqrt{(255^2 + 293^2)} = 389\,\text{mm}$$

Having calculated the length of cut on each slate, we must now determine the number of slates to be cut over a metre length of hip. This is done by first calculating the length of cut on the margin of

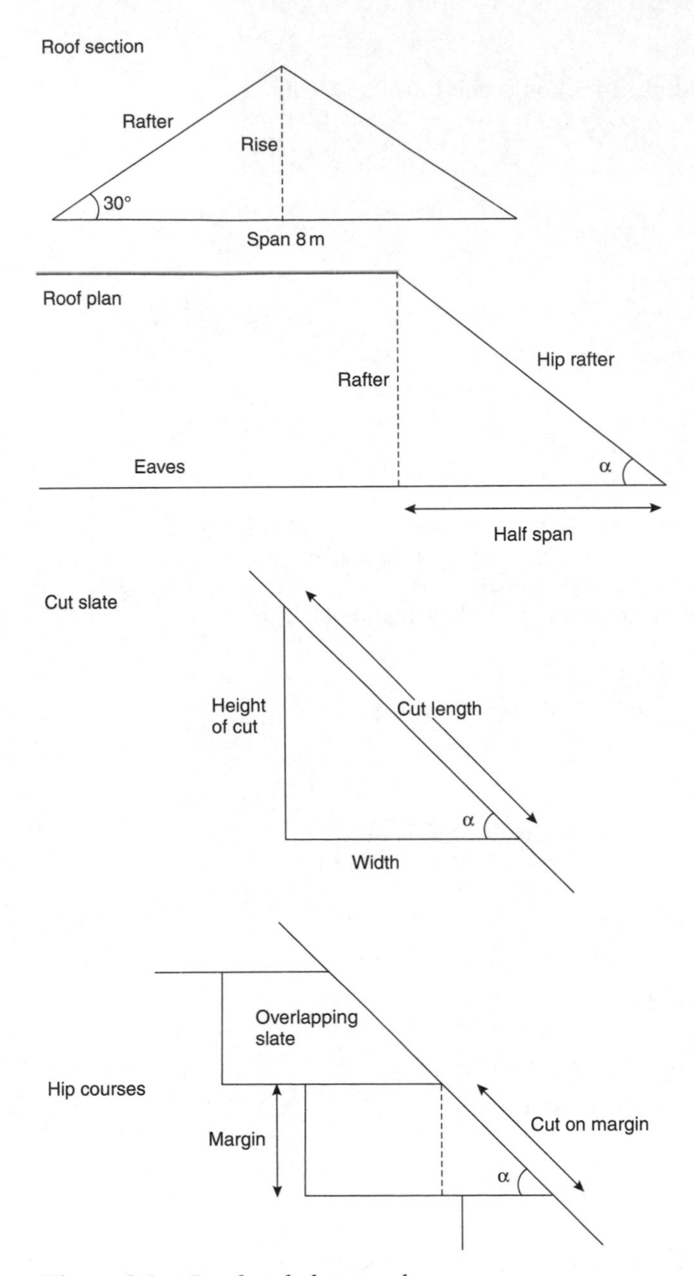

Figure 9.1 *Roof and slate angles*

each slate. With a plane angle of 49.11 degrees and a margin of 153 mm (as the gauge previously calculated), we firstly determine the length of cut on the margin as follows:

$$\sin \alpha = \frac{\text{margin}}{\text{cut on margin}}$$

$$\text{cut on margin} = \frac{\text{margin}}{\sin \alpha}$$

$$= \frac{153}{0.757}$$
$$= 202\,\text{mm}$$

We can now move on to estimation of costs.

Material: wastage
Nr courses slate per metre hip: 1000/202 = 4.95
Wastage on side of hip per metre: 4.95 × 50% = 2.48 slates

Cost of 1000 slates from Example 9.2		
Ex quarry	£780.00	
Delivery to site	£58.29	
Take delivery and stack	£9.75	
	£848.04	
Cost of slates per metre hip: £848.04 × (2.48/1000)		£2.10
Labour: cutting		
Cost squad per hour:		
Labour cost per hour (2 × £8.00) + (1 × £6.50) = £22.50		
Output per tradesman: 10 m hip cut per hour		
Cost to cut one metre:		£1.13
Cost to cut one side of hip		£3.23
Cost to cut two sides of hip		£6.46
Profit and oncost 20%		£1.29
Cost per metre		£7.75

Tile roofing

Roof tiles are manufactured from clay and concrete to a wide range of colours and profiles suitable for pitches from 17 to 45 degrees. Greater pitches up to vertical cladding are achievable but require additional or special fixing. Roof tiles are generally sold by the thousand in crates or strap banded to be stored on site pallets.

Clay plain roof tiles

British Standard BS 402 provides for tiles of nominal or work size 265 × 165 mm. Table 1 of the Standard gives maximum and minimum deviations for manufacturing size. Also detailed are requirements for tile manufacture, colour, nibs, thickness etc. Where tiles 280 × 175 mm are required for a project, but otherwise comply with BS 402, then such a requirement should be stated when ordering. BS 402 recognizes that plain tiles of different specification are manufactured, and although the Standard is not intended to provide for such tiles it does not rule out their use.

Plain tiles are laid using the double lap principle; this ensures that there will be at least two thicknesses of tile covering any part of the roof.

Table 9.4 *Range of tiler outputs fixing single-lap concrete tiles*

Overall work size(mm) Length	Width	Output per tradesman (m^3/hour)
381	229	3.50–4.00
413	330	3.25–3.75
420	330	3.25–3.75
430	380	3.00–3.50

Concrete roofing tiles

In accordance with the requirements of BS 473,550, concrete roofing tiles or concrete roofing slates are manufactured to be fixed either double lap or single lap. Single-lap tiles are fitted using interlocking principles; the lower edge of a tile is designed to fit closely to the upper edge of the tile adjacent. The overall work sizes and maximum and minimum deviations for manufacturing size are given in the Standard. The manufacture of tiles in sizes other than those stated in the Standard is not excluded, and such tiles may therefore be found in practice.

Table 9.4 shows typical overall work sizes for single-lap concrete tiles, and an indication of labour outputs based on a squad of two tradesmen to one labourer.

Concrete double-lap tiles have similar characteristics to those of clay plain tiles. The work sizes given in the Standard are 267 × 165 mm and 457 × 330 mm, and maximum and minimum deviations for manufacturing size are again specified.

Example 9.6

Roofing; concrete interlocking tiles 430 × 380 mm; fixing every tile with standard tile clip to 75 mm lap; battens 38 × 25 mm softwood sawn, pressure impregnated with preservative; on reinforced under-slating felt; coverings sloping to 40 degrees pitch.	m^2

Materials: tiles

Tiles delivered to site at £525.00 per 1000	£525.00
Take delivery and stack: 1.75 hours per 1000 at £6.50	£11.38
Clips: 1000 per 1000 tiles at £32.00 per 1000	£32.00
	£568.38
Allow waste 2.5%	£14.21
Cost per 1000 tiles ready for fixing	£582.59

Coverage 1 tile = gauge × exposed width
= (length − 75 mm lap) × (width − 35 mm lap)
= 0.355 × 0.345 = 0.12247 m^2

Coverage 1000 tiles: 0.12247 × 1000 = 122.48 m^2

Cost per m^2 = £582.59/122.48		£4.76
Underslating felt as Example 9.1: per m^2		£2.07
Materials: battens		
Nr of tiles per m^2: 1000/122.47 = 8.16		
Quantity 38 × 25 mm battens per m^2 = nr tiles per m^2 × exposed width		
= 8.16 × 0.345		
= 2.82 m		
Cost of battens per m^2: 2.82 m at £0.46 per m	£1.30	
Nr nails per m^2 = (length battens)/(rafter centres) = 2.82/0.45		
Weight nails per m^2 at 260 nails/kg: (2.82/0.45)/260 = 0.024 kg		
Cost of nails per m^2: 0.024 kg at £3.56/kg	£0.09	
	£1.39	
Allow waste 5%	£0.07	£1.46
Labour: battens		
Squad of 2 tradesmen and 1 labourer fixes 80 m battens per hour		
Labour cost per hour: £22.50		
Cost to fix battens for 1 m^2 slating: £22.50 × (2.82/80)		0.79
Labour: tiles		
From Table 9.4 for the tile size of 430 × 380 mm, 3.5 m^2 can be carried and fixed in position per hour per tradesman. There are 2 tradesman in the squad.		
Labour cost per hour = £22.50		
Cost per m^2 = £22.50/(2 × 3.5)		£3.21
		£12.29
Profit and oncost 20%		£2.46
Rate per m^2		£14.75

Example 9.7

> Roofing; plain clay rosemary tiles 265 × 165 mm; fixing every tile with one aluminium alloy nail, every fifth course twice nailed, to 65 mm lap; battens 38 × 25 mm softwood sawn, pressure impregnated with preservative; on reinforced underslating felt; coverings sloping to 40 degrees pitch. m^2

Materials: tiles	
Tiles delivered to site at £490.00 per 1000	£490.00
Take delivery and stack: 1.3 hours per 1000 at £6.50	£8.45
Nails: 38 mm aluminium at £5.98/kg at 1.75 kg per 1000: 1200 nails per 1000 slates	£12.56
	£511.01
Allow waste 2.5%	£12.78
Cost per 1000 tiles ready for fixing	£523.79

Coverage 1 tile = width × gauge
= 0.0165 × 0.1 = 0.0165 m^2

Coverage 1000 tiles : = $0.0165 \times 1000 = 16.5\,m^2$		
Cost per m^2 = £523.79/16.5		£31.74
Underslating felt as Example 9.1: per m^2		£2.07
Materials: battens		
Nr of tiles per m^2: 1000/16.5 = 60.61		
Quantity 38 × 25 mm battens per m^2 = nr tiles per m^2 × tile width		
= 60.61 × 0.165		
= 10 m		
Cost of battens per m^2: 10 m at £0.46 per m	£4.60	
Nr nails per m^2 = (length battens)/(rafter centres) = 2.82/0.45		
Weight nails per m^2 at 260 nails/kg: (10/0.6)/260 = 0.064 kg		
Cost of nails per m^2: 0.064 kg at £3.56/kg	£0.23	
	£4.83	
Allow waste 5%	£0.24	£5.07
Labour: battens		
Squad of 2 tradesmen and 1 labourer fixes 80 m battens per hour		
Labour cost per hour: £22.50		
Cost to fix battens for 1 m^2 slating: £22.50 × (10/80)		2.81
Labour: tiles		
For plain clay tiles 265 × 165 mm, 2.5 m^2 can be carried and fixed in position per hour per tradesman. There are 2 tradesman in the squad.		
Labour cost per hour = £22.50		
Cost per m^2 = £22.50/(2 × 2.5)		£4.50
		£46.19
Profit and oncost 20%		£9.24
Rate per m^2		£55.43

Example 9.8

Eaves; double course of 405 × 255 mm heather blue slates at eaves.	m

Materials: tiles	
Tiles delivered to site at £490.00 per 1000	£490.00
Take delivery and stack: 1.3 hours per 1000 at £6.50	£8.45
Nails: 38 mm aluminium at £5.98/kg at 1.75 kg per 1000: 2000 nails per 1000 slates	£20.93
	£519.38
Allow waste 2.5%	£12.98
Cost per 1000 tiles ready for fixing	£532.36
Nr tiles of width 165 mm per 100 m eaves: 100/0.165 = 606.06	
Cost of slates per 100 m eaves: £532.36 × (606.06/1000)	£322.64
Additional row of battens: 100 m at £0.46/m	£46.00

Nr nails = (length battens)/(rafter centres) = 100/0.6	
Weight of nails at 260 nails/kg: (100/0.6)/260 = 0.64 kg	
Cost of nails: 0.64 kg at £3.56/kg	£2.28

Labour: fixing battens and tiles
Squad of 2 tradesmen and 1 labourer fixes 80 m battens per hour

Labour cost per hour: £22.50	
Cost to fix 100 m battens: £22.50 × (10/80)	£2.81
Coverage 1 tile = width × gauge = 0.165 × 0.10 = 0.0165 m^2	
Coverage 1000tiles = 0.0165 × 1000 = 16.5 m^2	
Nr tiles per m^2 = 1000/16.5 = 60.61	
Area tiles per 100 m eaves = (nr tiles per 100 m eaves)/(nr tiles/m^2) = 606.06/60.61 = 10 m^2	
Labour cost £22.50/hour: output per tradesman 2.5 m^2/hour	
Cost to fix 100 m tiles at eaves: £22.50 × 10/(2 × 2.5)	£45.00
	£418.73
Profit and oncost 20%	£83.75
Cost per 100 m	£502.48
Rate per metre	£5.02

Example 9.9

Verges; single extra undercloak course 165 mm wide; bedding and pointing in cement mortar 1:3.	m

Materials

Tile-and-a-half tiles delivered to site at £930.00 per 1000		£930.00
Take delivery and stack: 1.35 hours per 1000 at £6.50		£8.78
Nails: 38 mm aluminium at £5.98/kg at 1.75 kg per 1000: 2000 nails per 1000 slates		£20.93
		£959.71
Allow waste 2.5%		£23.99
Cost per 1000 tiles ready for fixing		£983.70
Nr tiles of gauge 100 mm per 100 m verge: 100/0.1 = 1000		
Cost of tile-and-a-half tiles: 500 at £983.70 per 1000		£491.85
Coverage 265 × 165 mm tiles: 500 at £532.36 per 1000		£266.18
Cost of tiles per 100 mm verge		£758.03
Deduct cost of slates measured in superficial item (Example 9.7)		
Coverage tile-and-a-half tiles: 500 × (0.1 × 0.248)	12.40 m^2	
Coverage 265 × 165 mm tiles: 500 × (0.1 × 0.165)	8.25 m^2	
Area covered by verge	20.65 m^2	
Cost to deduct: 20.65 at £31.74 per m^2 (Example 9.7)		£655.43
Extra over cost of slates per 100 m verge		£102.60

Table 9.6 *Outputs in m²/hour for fixing lead sheet covering: squad of plumber and mate*

BS code	Flat roofs: <10° from horizontal	Roofs 10–50° from horizontal	Roofs 50° from horizontal to vertical
3	–	–	–
4*	0.90–1.00	1.00–1.10	1.40–1.55
5	1.15–1.25	1.30–1.40	1.70–1.80
6	1.35–1.45	1.50–1.60	2.05–2.20
7	1.55–1.65	1.70–1.85	2.30–2.45
8	1.80–1.90	2.00–2.20	2.60–2.80

* Code 4 lead sheet is suitable only for coverings to small areas for the types of work shown.

Underlay

The underlay used to isolate lead roofing, cladding and gutter linings from their substrates is impregnated felt or waterproof building paper. Impregnated felt should be used where the surface is not perfectly even. Where the surface is even, waterproof building paper to BS 1521 class A would be suitable.

Outputs

Ranges of squad outputs for lead sheet coverings are given in Table 9.6. The outputs include hoisting, cutting and fixing lead per square metre as measured. Ranges of squad outputs for lead flashings are given in Table 9.7. The outputs include hoisting, cutting and fixing lead per linear metre as measured.

Table 9.7 *Outputs in metres/hour for fixing lead sheet flashings: squad of plumber and mate*

BS code	Description	Girth (mm) <150	150–225	225–300
4	Flashing	0.25–0.30	0.35–0.40	0.40–0.45
5	Flashing	0.30–0.35	0.40–0.45	0.50–0.55
4	Stepped flashing	0.50–0.55	0.65–0.70	0.80–0.90
5	Stepped flashing	0.60–0.65	0.80–0.90	1.00–1.15
4	Hip or ridge capping	0.30–0.36	0.40–0.45	0.50–0.60
5	Hip or ridge capping	0.35–0.40	0.50–0.55	0.60–0.65
4	Sloping gutters	0.30–0.35	0.40–0.45	0.50–0.55
5	Sloping gutters	0.35–0.40	0.50–0.55	0.60–0.65
6	Sloping gutters	0.40–0.45	0.60–0.65	0.70–0.75

Example 9.11

Roof coverings; flat; Code 5 lead sheet on Erskine's felt underlay, fixing at head with copper clout nails 25 mm long 10 SWG, copper side edge and end retaining clips; wood core rolls at 1000 mm centres; drips every 2000 mm.	m^2

Materials		
Cost of 1000 kg of lead sheet	£865.00	
Plumber's mate taking delivery at 1000 kg/hour at	£6.50	
Cost per 1000 kg	£871.50	
Cost per m^2 of code 5: £871.50 × (25.4/1000)		£22.14
Additional 250 mm width of lead required forming laps over 45 mm wood roll:		
Cost for 250 mm lap per m^2: £22.14 × (250/1000)		£5.54
Additional 180 mm length of lead required forming drips every 2 m:		
Cost for 180 mm lap per m^2: £22.14 × (180/1000)		£3.99
Copper clout nails per m^2: say 31 at 120/kg at £7.35/kg		£1.90
Copper clips at 500 mm centres fixed with 2 nails/clip (nails allowed for above): 3 clips per m^2 at £0.05		£0.15
Cost per m^2 for underlay including allowance for laps and drips: say		£1.67
		£35.39
Allow for waste 2.5%		£0.88
Labour		
Squad will hoist, cut and fix 1 m^2 in 1.2 hours:		
Plumber: 1.2 hours at £9.30		£11.16
Plumber's mate: 1.2 hours at £7.10		£8.52
		£55.95
Profit and oncost 20%		£11.19
Rate per m^2		£67.14

Example 9.12

Flashings; horizontal; Code 5 lead sheet; turning into groove and fixing with milled lead wedges at 200 mm centres; 150 mm girth.	m

Materials		
Cost of 1000 kg of lead sheet	£865.00	
Plumber's mate taking delivery at 1000 kg/hour at	£6.50	
Cost per 1000 kg	£871.50	
Cost per m^2 of Code 5: £871.50 × (20.41/1000)	£17.79	
Cost per metre at 150 mm girth: £17.79 × 0.15		£2.67
Additional 100 mm width of lead required forming laps every 2 m:		
Cost for 100 mm lap per metre: £2.67 × 100/(2 × 1000)		£0.13

Wedges 5 per metre: say	£0.36
	£3.16
Allow for waste 2.5%	£0.08
Labour	
Squad will hoist, cut and fix 1 m in 0.28 hours:	
Plumber: 0.28 hours at £9.30	£2.60
Plumber's mate: 0.28 hours at £7.10	£1.99
	£7.83
Profit and oncost 20%	£1.57
Rate per metre	£9.40

Example 9.13

Gutters; sloping; Code 5 lead sheet; dressing over tilting fillets; fixing with copper clout nails; 550 mm girth.	m

Materials		
Cost of 1000 kg of lead sheet	£865.00	
Plumber's mate taking delivery at 1000 kg/hour at	£6.50	
Cost per 1000 kg	£871.50	
Cost per m^2 of Code 5: £871.50 × (25.4/1000)	£22.14	
Cost per metre at 550 mm girth: £22.14 × 0.55		£12.18
Additional 100 mm width of lead required forming laps every 2 m:		
Cost for 100 mm lap per metre: 12.18 × 100/(2 × 1000)		£0.61
Copper clout nails per m^2: say 15 at 120/kg at £7.35/kg		£1.90
		£14.69
Allow for waste 2.5%		£0.37
Labour		
Squad will hoist, cut and fix 1 m in 0.73 hours:		
Plumber: 0.73 hours at £9.30		£6.79
Plumber's mate: 0.73 hours at £7.10		£5.18
		£27.02
Profit and oncost 20%		£5.40
Rate per metre		£32.42

Example 9.14

Edges; welted.	m

Plumber forms 1 m of welted edge on Code 5 lead in 0.12 hours:

Plumber: 0.12 hours at £9.30	£1.12
	£1.12
Profit and oncost 20%	£0.22
Rate per metre	£1.34

10

Asphalt work

SMM7 Sections J20, J21, M11

Introduction

It is perhaps unfair to refer to any particular trade as special, for every operative in the industry demonstrates skills which are unique to the craft he follows. However, asphalt work is generally referred to as a specialist trade because it encompasses some unique features and is practised by a relatively small number of craftsmen employed by a relatively small number of companies. To further emphasize this specialism, contracts for this type of work are normally let as a subcontract and most commonly to a nominated subcontractor.

The most unique aspect of the trade is the material used. Unlike every other trade, the material, mastic asphalt, is handled by the operatives while it is hot, generally between 200 and 220 °C. Pitch mastic is laid at the lower temperature of 160 °C.

This has two consequences. The first is that the work is dangerous; protective clothing is essential and safety procedures must be adhered to.

The second is that a relatively rapid pace is needed in the application of the material, as it sets by cooling to ambient temperature. As soon as the material is drawn from the cauldron or mixer it starts to cool. When it is applied to a surface the rate of cooling accelerates for two reasons. First, the base surface is at ambient temperature and a base of masonry presents an enormous heat sink. Second, as the material is spread out, more of it comes into contact with this cold base and with the cool air over it.

Perhaps as a consequence of the specialist nature of the work there seems to be no well-defined procedure used by estimators for building up rates for measured items of work. Therefore the data and examples which follow are given only as a guide to one possible approach to achieve comprehensive financial cover for all expenditure.

Before commencing examples it is proper to consider the logistics and practice in the trade.

Plant

Plant can be divided between melting equipment and small tools.

Melting equipment comes in a range of capacities, that is the weight of asphalt or pitch with which it is possible to deal in one operation. The smaller capacity items are called cauldrons or pots. Pots comprise a cylindrical lidded container with a heat source underneath. These pots are generally fired from cylinders of propane gas. While mastic asphalt is being melted and kept molten it is occasionally stirred to keep the temperature even. Pitch mastic has to be continuously stirred, and pots for use with that material have mechanically driven paddles built in for that purpose. Pots generally have a capacity of 500 kg of asphalt; this will take approximately 25 kg of propane to melt down.

Larger capacity plants are called mixers. The melting chamber incorporates paddles for mechanical mixing. The fuel used was once coke or coal but this has been replaced with oil or propane. Mixers may be mounted permanently on either a lorry or a trailer chassis. Usually mixers have a capacity of 1500 kg of asphalt, and naturally take proportionately more fuel to melt down all the asphalt.

Small tools and accessories include the following:

- Various trowels for spreading asphalt, working fillets, turning into grooves, etc.
- Lidded buckets for transporting the melted asphalt from pot to site of laying
- Rules, straight edges and gauges
- Tin snips, pliers, hammers, knives, etc. to deal with underlays and reinforcing meshes.

Materials

Rock and mastic asphalt and pitch mastic are purchased by weight and supplied in blocks or cakes, usually weighing 25 kg.

Typically 1 tonne (1000 kg) of asphalt will cover 35 m^2 to a thickness of 12 mm, this being the thinnest single coat usually laid in one-coat work. Coats as thin as 10 mm are laid but only in combination with other coats. This cover rate takes into account a certain degree of unevenness in the substrate. The figure of 12 mm therefore would be used for a first coat. Subsequent coats will have an increased cover rate as there is no need to take any unevenness into account; the increase in cover rate is approximately 5 per cent.

So for a layer 20 mm thick, whatever the number of coats, the quantity of asphalt used per m^2 would be as follows:

First 12 mm thickness: 1000/35	28.57 kg
Further 8 mm thickness: (1000/35) × (8/12) × 1.05	20.00 kg
	48.57 kg

This is equivalent to 1000/48.57 = 20.59 m^2 per tonne. Waste is not included in the calculation; it is normally added to the cost of the material when working out the rate (see later). Bituminous felt and glass fibre cloth or tissue are used where an underlay is required; the bituminous felt should be type 4A(i) complying with BS 747. Bitumen coated expanded metal lath is used as a reinforcement for vertical work and in upstands, where adhesion to the substrate is not possible; this is supplied in sheets of approximately 2500 × 700 mm. Galvanized clout nails or wire staples are used to secure underlay and reinforcement to timber substrates.

Rubber/bitumen primer is used to coat metallic surfaces prior to the application of asphalt. This ensures a watertight bond. Coverage is approximately 10 m^2 per litre. Surfaces treated might include lead flashings and drips, cast iron rainwater outlets and gullies in roofs and pavings, pipes passing through a surface, balusters and other metal work. The primer has a solvent base – usually petroleum spirit – and is highly flammable in its liquid state.

Sundry materials used include crushed limestone dust to line buckets, thus preventing the asphalt or pitch mastic sticking to the bucket, and sharp sand to obtain a sand rubbed finish.

Labour

The size and nature of the job, the timetable and the accessibility of the works will determine whether pots or mixers are to be used. There may be a combination of a mixer for the large areas and a pot for the ancillary work such as skirtings, upstands, kerbs and drips.

The squad make-up and cost are generally broken down as follows:

Squad Cost

	Number	Rate	Costs
Craftsman (spreader)	2	£8.00	£16.00
Potman	1	£6.85	£6.85
Labourer	1	£6.50	£6.50
Squad cost per hour			£29.35
Cost per craftsman per hour			£14.68

Table 10.1 *Range of outputs per craftsman for mastic asphalt work*

(a) Area outputs: m^2/hour

Thickness (mm)	12–13	15–17	18–20
Coats	1	2	2
Laying flooring/paving			
Horizontal/falls	0.24–0.43	0.29–0.46	0.32–0.49
Laying roofing			
Horizontal	0.30–0.60	0.33–0.75	0.50–0.90
Sloping	0.40–0.72	0.60–1.08	0.90–1.38
Vertical	1.00–0.56	1.20–2.16	1.85–2.40

(b) Linear and enumerated items (m/hour and hours each)

	mm	hours
Skirtings, aprons, upstands by girth	150	0.36–1.00
	225	0.44–1.32
	300	0.50–1.65
Angles by girth	150	0.45–0.60
	225	0.65–0.80
	300	0.85–1.00
Forming channels in asphalt by girth	225	0.70–0.90
	300	1.00–1.20
	450	1.25–1.40
Rounded edges		0.20–0.33
Arris: horizontal		0.30–0.50
vertical		0.25–0.40
Internal angle fillet: horizontal		0.15
vertical		0.20
Working into groove: horizontal		0.30
vertical		0.35

The output for the squad is based on how much material a craftsman can lay per squad hour. The output of the craftsman is affected by the nature of the work and its location. For example, work may be in a basement, in an underpinning, on a roof or on a floor; it may be in small areas or confined spaces; it may be sloping or vertical; and it may require to be done overhand. Typical outputs are given in Table 10.1. Rule 1.*.*.3 of SMM7 Section J 'Waterproofing' is generally only applicable to cold applied materials.

Wastage

Generally 2.5–5 per cent is allowed on asphalt and underlay. Small amounts of reinforcement and the priming of numerous small items will incur a greater degree of wastage in those particular items. However, the difference is small enough to be ignored and a general 5 per cent is used throughout the examples here.

Building-up rates

Before commencing examples of building-up rates, let us establish a charge for the major plant item – the pot – and then set a small scenario within which to price a few items.

The hire rate for a half tonne pot for one week is £47.00. For a 39 hour week, the hire is therefore £1.22 per hour.

The total area of asphalt to be laid is taken to be 230 m^2 plus sundry skirtings, angle fillets, drips, etc. Generally the thickness is 20 mm. Thus, using the figure calculated earlier, the approximate quantity of asphalt is 230 × 48.57 kg, plus 10 per cent for other items: say 12.29 tonnes. A close approximation of propane used is 1 kg per 10 kg of mastic asphalt. We will take 1 kg propane gas at a cost of £0.92.

Example 10.1

20 mm thick mastic asphalt in two coats to horizontal surface.	m^2

Material:	
Mastic asphalt 48.57 kg at £98/tonne	£4.76
Type 4A(i) sheathing felt as isolating membrane:	
Roll 20 m^2 costs £12.50: divide by 19 m^2 to allow for laps	£0.66
	£5.42
Waste 5%	£0.27
	£5.69
Plant:	
500 kg asphalt is melted in 1.5 hours	
To melt 48.57 kg: hire charge (£1.22 × 1.5 × 48.57)/500	£0.18
Transport pot to and from site: £30 shared over 12.29 tonnes	
So for 48.57 kg say (£30.00 × 48.57)/(12.29 × 1000)	£0.12
Propane to melt 500 kg, 50 kg at £0.92	
So to melt 48.57 kg will cost (£0.92 × 50 × 48.57)/500	£4.49
	£10.47
Labour laying 3 m^2/hour at £14.68	£4.89

Pot hire charge while laying hourly cost/3 m^2	£0.41
	£15.77
Profit and oncost 20%	£3.15
Rate per m^2	£18.92

Example 10.2

20 mm thick mastic asphalt in two coats to horizontal timber substrate, including isolating membrane nailed to timber deck.	m^2

Material: 48.57 kg at £98/tonne	£4.76
Type 4A(i) sheathing felt as isolating membrane:	
Roll 20 m^2 costs £12.50: divide by 19 m^2 to allow for laps	£0.66
Clout nails 150 mm centres each way: 0.10 kg at £3.40/kg	£0.34
	£5.76
Waste 5%	£1.15
	£6.91
500 kg asphalt is melted in 1.5 hours	
To melt 48.57 kg: hire charge (£1.22 × 1.5 × 48.57)/500	£0.18
Transport pot to and from site: £30 shared over 12.29 tonnes	
So for 48.57 kg say (£30.00 × 48.57)/(12.29 × 1000)	£0.12
Propane to melt 500 kg, 50 kg at £0.92	
So to melt 48.57 kg will cost (£0.92 × 50 × 48.57)/500	£4.49
	£11.69
Labour laying 3 m^2/hour at £14.68	£4.89
Pot hire charge while laying hourly cost/3 m^2	£0.41
	£16.99
Profit and oncost 20%	£3.40
Rate per m^2	£20.39

Example 10.3

20 mm thick mastic asphalt in two coats to vertical face of brickwork.	m^2

Mastic asphalt 48.57 kg at £98/tonne	£4.76
Waste 5%	£0.95
	£5.71
500 kg asphalt is melted in 1.5 hours	
To melt 48.57 kg: hire charge (£1.22 × 1.5 × 48.57)/500	£0.18
Transport pot to and from site: £30 shared over 12.29 tonnes	
So for 48.57 kg say (£30.00 × 48.57)/(12.29 × 1000)	£0.12
Propane to melt 500 kg, 50 kg at £0.92	

So to melt 48.57 kg will cost (£0.92 × 50 × 48.57)/500		£4.49
		£10.50
Labour laying 0.4 m²/hour at £14.68		£36.69
Pot hire charge while laying hourly cost/0.4 m²		£3.04
		£50.23
Profit and oncost 20%		£10.05
Rate per m²		£60.27

Example 10.4

20 mm thick mastic asphalt in two coats in skirting 150 mm high including angle fillet, rounded arris and tucking into groove in brickwork.	m

Volume of material in skirting (m³/m):		
Main upstand 150 × 20 mm	0.0030	
Angle fillet 50 mm wide across angle: (50 × sin 45°) = 35.36 mm side	0.0006	
Groove in brickwork	0.0005	
Total volume	0.0041	
Density of mastic asphalt 2381 kg/m³		
Material: 2381 × 0.0041 = 9.76 kg at £98/tonne		£0.96
Waste 5%		£0.05
		£1.00
500 kg asphalt is melted in 1.5 hours		
To melt 9.76 kg: hire charge (£1.22 × 1.5 × 48.57)/500		£0.04
Transport pot to and from site: £30 shared over 12.29 tonnes		
So for 9.76 kg say (30.00 × 9.76)/(12.29 × 1000)		£0.02
Propane to melt 500 kg, 50 kg at £0.92		
So to melt 9.76 kg will cost (£0.92 × 50 × 9.76)/500		£0.90
		£1.97
Labour laying 2 m/hour at £14.68		£7.34
Pot hire charge while laying hourly cost/2 m		£0.61
		£9.91
Profit and oncost 20%		£1.98
Rate per metre		£11.89

Example 10.5

20 mm thick mastic asphalt in two coats in skirting 150 mm high, on free-standing timber kerb, including angle fillet reinforcement and isolating membrane.	m

Volume of material in skirting (m³/m):		
Main upstand 160 × 20 mm (160 girth over timber)	0.0032	

Angle fillet 50 mm wide across angle: (50 × sin 45°) = 35.36 mm side	0.0006	
Total volume	0.0038	
Density of asphalt 2381 kg/m³		
Material: 2381 × 0.0038 = 9.05 kg at £98/tonne		£0.89
Reinforcement: 225 mm wide at £104/m²		£0.32
Type 4A(i) sheathing felt: 150 mm wide at £0.53/m²		£0.08
Nails, say		£0.04
		£1.32
Waste 5%		£0.07
		£1.39
500 kg asphalt is melted in 1.5 hours		
To melt 9.05 kg: hire charge at £1.22/hour		£1.10
Transport pot to and from site: £30 shared over 12.29 tonnes		£2.49
So for 9.05 kg say (£30.00 × 9.05)/(12.29 × 1000)		£0.02
Propane to melt 500 kg, 50 kg at £0.92/kg		
So to melt 9.05 kg will cost (£0.92 × 50 × 9.05)/500		£0.84
Labour laying 1.8 m/hour at £14.68		£8.15
Pot hire charge while laying hourly cost/1.8 m		£0.68
		£12.18
Profit and oncost 20%		£2.44
Rate per metre		£14.61

Example 10.6

Internal angle fillet.	m

Volume of material in fillet (m³):		
Angle fillet 50 mm wide across angles: (50 × sin 45°) = 35.36 mm side	0.0006	
Total volume	0.0006	
Density of asphalt 2381 kg/m³		
Material: 2381 × 0.0006 = 1.43 kg at £98/tonne		£0.14
Waste 5%		£0.01
		£0.15
500 kg asphalt is melted in 1.5 hours		
To melt 1.43 kg: hire charge at £1.22/hour		£0.17
Transport pot to and from site: £30 shared over 12.29 tonnes		
So for 1.43 kg say (£30.00 × 1.43)/(12.29 × 1000)		£0.003
Propane to melt 500 kg, 50 kg at £0.92/kg		
So to melt 9.05 kg will cost (£0.92 × 50 × 9.05)/500		£0.13
		£0.46

Labour laying 6 m/hour at £14.68	£2.45
Pot hire charge while laying hourly cost/1.8 m	£0.20
	£3.24
Profit and oncost 20%	£0.65
Rate per metre	£3.89

Example 10.7

Forming channel girth 300 mm in asphalt.	m

Labour forming 1 m/hour at £14.68	£14.68
Profit and oncost 20%	£2.94
Rate per metre	£17.62

11

Woodwork carpentry: first fixings

SMM7 Sections G20, K10, K31

The rationale behind the layout and structure of the rules of measurement in SMM7 has done away with the work section/trade headings which have been so familiar in the past. The versatility of wood and wood products and of the carpenters and joiners themselves now means that the rules of measurement for their work are well scattered through the new document. However, in this and the following chapter a selection of items and rates for these traditional trade divisions is presented in an order similar to that of SMM7.

Special requirements

In dealing with any woodworking items the estimator will be on the lookout for the following information, which will denote something additional or unusual and therefore possibly expensive!

Species of timber

Hardwoods are generally more expensive than softwoods, and in either classification there is a wide range of cost due to what might appear to be a diversity of factors. These, however, generally all boil down to scarcity of a suitable supply of timber. It is also true to say that timbers are more expensive when required in larger cross-sections and/or long continuous lengths. These latter factors can be overcome using machine jointing and modern synthetic glues; the joint is often stronger than the raw timber but the jointing process does add to the cost. However, if timbers in excess of 5.4 m long are required this will add around 12.5 per cent to the basic cost of timber. In excess of 7.5 m long an addition of 25 per cent or more would be quite likely.

Stress grading

This is an added expense but one which may save money in the long run by allowing the use of smaller sections of timber. Stress grading adds in excess of 7.5 per cent for GS grade timber to over 20 per cent for SS grade timber to the basic cost of timber; however, the cross-section of the timber also affects the cost on a modestly sliding scale.

Grade of sheet material

There are many different board and sheet materials on the market. A large number of them might be described as, say, plywood. However, there is a vast difference in price between a sheathing plywood, a Dougals Fir plywood and a hardwood veneered plywood for decorative work, or between one where the glue is graded water and boil proof (WBP) and another graded moisture resistant (MR).

Similarly, there are many different types and grades of particle board, lamin board, blockboard and so on.

Pre-treatment

This covers treatment with chemicals against insect or fungal attack or for increased fire resistance. The treatment may be applied with a simple brush, in vacuum/pressure plant, etc. Note that sheet material can be pre-treated as well as natural timber. Currently pressure impregnation against fungal or insect attack with copper chrome arsenate (CCA) salts-based preservative adds around £20–£25 per cubic metre (second quarter of 2002) to the basic cost of timber. The smaller the cross-section of the timber treated, the greater the amount of preservative fluid used and so the higher the costs.

Machined timber

Timber may be supplied regularized or wrot rather than off the saw, or may be sawn to a special shape.

It is impossible to say what percentage to add to the basic timber cost for regularizing or machine planing, since the act of machining the surface of a rectangular shape is much the same whether the cross-section is 38 × 38 mm or 100 × 100 mm. What does vary of course is the proportion of waste produced to the remaining processed timber. For a section 38 × 38 mm (1444 mm^2), fully wrot surfaces will mean that the timber will end up as 32 × 32 mm (1024 mm^2) maximum. This represents a loss of 420 mm^2 or 29.09 per cent. The 100 × 100 mm timber (10 000 mm^2) will end up as 94 × 94 mm (8836 mm^2). This represents a loss of 1164 mm^2 or 11.64 per cent. If the timber is 200 × 200 mm, the loss will be 5.91 per cent. The following are typical costs of these timbers as quoted by timber suppliers in the second quarter of 2002

	Cost per 100 m	
Basic size (mm)	Sawn	Wrot
38 × 38	£35	£57
100 × 100	£217	£290
200 × 200	£810	£870

Examples of special shapes produced by sawing are arris, cant and firring. These add around 15 per cent to the cost of the basic timber.

Timber sizes and costs

The raw material, whether for direct use on site or for prefabrication, may be purchased by the metre. Usually the price quoted is for 100 m of a specific cross-section. More rarely timber is purchased by the cubic metre, but generally this is by the shipload and without conversion or processing between unloading and delivery.

Timber is supplied in a range of cross-sections. The following common names are approximately in descending order of cross-sectional area: baulks, half-timbers, flitches, planks, deals, boards, strips, battens, squares, slating battens and scantlings.

Timber is also supplied in one of four conditions: sawn or off the saw; regularized or surfaced, i.e. machined to a constant cross-section; wrot or dressed, i.e. machined to a constant cross-section and a smooth finish suitable for joinery work; and moulded, i.e. wrot and with a shaped cross-section.

BS 4471, 'Sizes of sawn and processed softwood', gives the sawn sizes of softwoods together with the reduction in overall size when the timbers are regularized or wrot. Note that flooring boards are treated differently and are the subject of BS 1297, 'Grading and sizing of softwood flooring'. Material from which UK sawmills machine flooring boards is specifically sawn for that purpose at the original sawmill before being shipped to the UK; the contractor buys the machined boards. BS 5450, 'Sizes of hardwoods', gives sawn sizes and processing allowances for planing, etc.

As well as standard cross-sectional sizes, softwoods are supplied in standard lengths. The shortest length normally supplied is 1.8 m. Standard lengths increment by 300 mm from this basic length. The range of sizes continues therefore as 2.1, 2.4, 2.7, 3.0 m, etc.

If these lengths do not fit the work required of them exactly, there is a maximum wastage of 299 mm to cut the next largest size down. The bulk of timber is purchased in the range of lengths from 3 to 6 m. If the wastage of 299 m is taken on the average length of 4.5, it is equivalent to 6.64 per cent. Many authorities quote 7.5 per cent allowance for wastage, which has also to include pilferage and damage. In the examples here we will take 10 per cent on timber, 7.5 per cent on manufactured units such as trusses, and 2.5 per cent on manufactured units such as doors and windows.

Although supplied in such a wide variety of sizes, natural timber is generally priced per cubic metre for a range of sizes and finishes. Pre-treatment is also costed out on a cubic metre basis. Prefabricated work (manufactured units) is generally purchased on a unit basis, i.e. per door, per flight of stairs, per kitchen unit, per trussed rafter, etc.

As in all the other trades. the estimator must ensure that delivery costs to site are included in any quotes or must make an appropriate allowance. Unloading on site is most often done by having a mechanical off-load (MOL) facility on the delivery truck, or by having rough terrain fork-lifts to cope with banded or palleted materials. A fork-lift would be priced in the preliminaries section.

Fastenings and adhesives

Fastenings and adhesives are not generally measured separately. Nails are deemed to be included; all other fastenings are given in the item description. There is a wide range of fastenings available, each suited to a particular application.

Nails can be manufactured with round flat heads or bullet heads, and from round or oval wire. The materials range through mild steel, stainless steel, brass, copper, aluminium, galvanized steel, zinc plated steel, sherardized steel, etc. Shanks can be plain or jagged, or can have an annular ring barb to improve holding power. An old form of nail which was punched out of steel sheet has enjoyed a revival in recent decades, as it has been found suitable for direct fastening into autoclaved aerated concrete blockwork and no-fines concrete. These nails were known as cut nails and were produced in a variety of patterns, one of which was specifically designed for fastening flooring boards. Others were designed for anything from general carcassing to fine cabinet and upholstery work. Nails are well illustrated in BS 1202. Wood screws are manufactured in a similar wide range of materials and finishes, and with a variety of heads (countersunk, round, pan, etc. as well as slotted, Phillips, Posidrive, etc.) and thread forms. Wood screws are fully specified in BS 1210. In addition there is a range of bolts, lag screws, fastenings for attachment to masonry, patent clips, hangers, hold-down straps, etc. far too numerous to mention here. Manufacturers' catalogues give details of proprietary fastenings.

Glues can have several different prime sources, including animal bones, hooves and horns, fish, milk, cellulose and various synthetic rubbers and resins. Glues based on animal, milk and cellulose products are generally classified as being from unmodified natural materials.

Since the late 1940s there has been a rapid growth in products based mainly on synthetic materials, and these have provided the industry with adhesives which cover every requirement. BS 5442:Part 3 gives guidance on the suitability of all the glues mentioned above for use with wood. The synthetic materials are classified as synthetic thermoset resins, elastomers and thermoplastics.

The carpenter and joiner is expected to fix items of ironmongery ranging from simple hat and coat hooks through complex items securing doors and windows, patent fasteners and openers for sashes and doors, to control gear for large industrial doors. The range of ironmongery is too extensive to list here, and manufacturers' catalogues and instructions should be consulted.

Round wire nails in plain steel are the most common method of fastening for general carcassing work. Galvanized round wire nails are used when simple rusting would be a problem. Under the rules of measurement, plain steel nails are the norm and are deemed to be included. Galvanized nails have to be given in the item description, the bill headings or the preamble. Similar provisions apply for nails in other materials and finishes, and for other more complex (and more expensive) forms of fasteners.

Labour constants

In general terms, hardwood is more difficult to work with than softwood. Various authorities quote labour constants for softwood, and show multipliers ranging from 1.5 to 2.5 to be applied to these constants for working with hardwood in general. Others give multipliers for specified hardwoods, recognizing that not all hardwoods have the same physical hardness or ease of cutting and machining.

Of course neither are all softwoods the same. Some of the so-called softwoods are harder and heavier and more difficult to machine than the majority of hardwoods; for example, pitch pine (*Pinus pulustris*) is quite dense and very resinous, which makes it difficult to work with hand tools. Modern construction practice tends to utilize pines, spruces and firs specially grown or selected from natural stands for their easy working capabilities and a general uniformity of quality. This obviates any need to differentiate between varieties as far as labour constants are concerned. The estimator must beware, however, of the occasional use of some unusual material in carpentry work. Many Far Eastern and South American exporters are promoting the use of unusual hardwoods for carpentry work in the face of an environmental movement in Europe concerned about the felling of rain forests!

Table 11.1 gives a range of labour constants per linear metre for various carpentry timbers based on their cross-sectional area. Table 11.2 gives a further range of constants per cubic metre for general classifications of carpentry timber(s). Both tables assume general softwood – pines, spruces and firs.

Table 11.2 *must be treated with caution*. It is not feasible to convert these labour hours for 1 m^3 directly into hours per metre for a particular cross-section of joist, plate, etc. Table 11.3 shows the results of such a conversion for sleeper joists using a figure of 20 hours/m^3. Comparison with the figures in Table 11.1 shows how the labour for floor timbers has been **overestimated**. Similarly, Table 11.4 shows calculations for roof timbers on the basis of 30 hours/m^3. Again these **overestimate** the constants per metre.

So why bother with hours for 1 m^3 of timber? Well, it is convenient for pricing works where there are no detailed quantities and no time to produce them. The inaccuracies will even out if the job is large enough, but the total result can only be an approximate estimate and not an accurate prediction of cost. Therefore costings based on m^3 of timber must be treated with care. In the examples here, constants based on individual cross-sectional areas will be used.

Table 11.1 *Range of labour constants for fixing selected carcassing timbers (per carpenter)*

Item	Floor/roof joists/ties hours/m	Other roof timbers hours/m
Wall plates (any size)	0.04–0.13	0.04–0.13
Other: 100 × 50, 125 × 50	0.05–0.17	0.10 0.18
150 × 50, 175 × 50	0.07–0.17	0.14–0.20
200 × 50, 225 × 50	0.10–0.19	0.18–0.24
Strutting: solid	0.33–0.40	
Herringbone	0.30–0.50	
Nails	0.15–2.00 kg/m^3	
Trimming openings:		
Trim and house one timber		0.20–0.40 hours each
Tusk tenon and house one timber		1.00–2.00 hours each

Table 11.2 *Labour constants for fixing 1 m^3 of carcassing timbers (per carpenter)*

Item	hours/m^3
Wall plate	7–12
Any floor or ceiling joist	8–20
Other roof timbers	20–30
Nails	1.1–4 kg/m^3

Table 11.3 *Conversion of a labour constant of 20 hours/m^3 to constants per metre floor joists (per carpenter)*

First size (mm)		Second size (mm)	m/hour	hours/m
50	×	100	10.00	0.100
50	×	125	8.00	0.125
50	×	150	6.67	0.150
50	×	175	5.71	0.175
50	×	200	5.00	0.200
50	×	225	4.44	0.225

Table 11.4 *Conversion of a labour constant of 30 hours/m³ to constants per metre for roof timbers (per carpenter)*

First size (mm)		Second size (mm)	m/hour	hours/m
50	×	100	15.00	0.150
50	×	125	12.00	0.188
50	×	150	10.01	0.225
50	×	175	8.56	0.263
50	×	200	7.50	0.300
50	×	225	6.66	0.338

Examples of rates calculation

It should be explained that the use of labourers by carpenters is not common and depends very much on the custom and size of the company in which they work. For the purposes of the following examples, a squad of five carpenters and one labourer is assumed. The squad cost is broken down as follows:

1 labourer at £6.50	£6.50
5 carpenters at £8.00	£40.00
Squad cost per hour	£46.50
Cost per carpenter hour	£9.30

It is further assumed that timber and components delivered to site are banded or palleted for mechanical off-loading. Off-loading is therefore included elsewhere in the pricing of the bill of quantities.

All timbers are assumed to be untreated whitewood, which is sawn, regularized or wrot as appropriate.

Example 11.1

Gang nailed trussed rafter, 35° pitch and 8 m span; timber assumed to be regularized. Nr

Trussed rafter delivered to site		£38.00
Unload and stack: 0.1 hours at £6.50		£0.65
		£38.65
Waste 5%		£1.93
Squad hoists and sets: 0.4 carpenter hours at £9.30		£3.72
		£44.30
Nails: 0.05 kg at £1.02/kg	£0.05	

Waste 10%	£0.01	£0.06
		£44.36
Profit and oncost 20%		£8.87
Rate per truss		£53.23

Example 11.2

Whitewood, sawn sleeper joists, laid on plates.	m

	150 × 38 mm	100 × 50 mm	125 × 50 mm	150 × 50 mm
Material cost per 100 m	£127.00	£102.00	£127.00	£150.00
Nails: 1 kg at £1.02	£1.02	£1.02	£1.02	£1.02
	£128.02	£103.02	£128.02	£151.02
Waste 10%	£12.80	£10.30	£12.80	£15.10
	£140.82	£113.32	£140.82	£166.12
Squad laying varying quantities per carpenter hour at £9.30				
150 × 38 mm: 24 m	£38.75			
100 × 50 mm: 24 m		£38.75		
125 × 50 mm: 19 m			£48.95	
150 × 50 mm: 16 m				£58.13
	£179.57	£152.07	£189.77	£224.25
Profit and oncost 20%	£35.91	£30.41	£37.95	£44.85
	£215.49	£182.49	£227.72	£269.10
Rate per metre	£2.15	£1.82	£2.28	£2.69

Example 11.3

Whitewood, sawn upper floor joists, laid on plates.	m

	150 × 38 mm	100 × 50 mm	125 × 50 mm	150 × 50 mm
Material cost per 100 m	£127.00	£102.00	£127.00	£150.00
Nails: 1 kg at £1.02	£1.02	£1.02	£1.02	£1.02
	£128.02	£103.02	£128.02	£151.02
Waste 10%	£12.80	£10.30	£12.80	£15.10
	£140.82	£113.32	£140.82	£166.12

Squad laying varying quantities per carpenter hour at £9.30

150 × 38 mm: 10.5 m	£88.57			
100 × 50 mm: 10 m		£93.00		
125 × 50 mm: 9.15 m			£101.64	
150 × 50 mm: 8 m				£116.25
	£229.39	£206.32	£242.46	£282.37
Profit and oncost 20%	£45.88	£41.26	£48.49	£56.47
	£275.27	£247.59	£290.95	£338.85
Rate per metre	£2.75	£2.48	£2.91	£3.39

Example 11.4

100 × 38 mm redwood sawn Tanalized wall plate handed to builder for bedding.	m

Material cost per 100 m	£105.00
Labour handling: 0.3 hours at £6.50	£1.95
	£106.95
Profit and oncost 20%	£21.39
	£128.34
Rate per metre	£1.28

Example 11.5

Whitewood, sawn ceiling joists and flat roof joists, laid on plates; and rafters, collars/struts in stick built roofs.	m

	150 × 38 mm	100 × 50 mm	125 × 50 mm	150 × 50 mm
Material cost per 100 m	£127.00	£102.00	£127.00	£150.00
Nails: 1 kg at £1.02	£1.02	£1.02	£1.02	£1.02
	£128.02	£103.02	£128.02	£151.02
Waste 10%	£12.80	£10.30	£12.80	£15.10
	£140.82	£113.32	£140.82	£166.12
Squad laying varying quantities per carpenter hour at £9.30:				
150 × 38 mm: 16 m	£58.13			
100 × 50 mm: 16 m		£58.13		
125 × 50 mm: 13 m			£71.54	
150 × 50 mm: 11 m				£84.55
	£198.95	£171.45	£212.36	£250.67

Profit and oncost 20%	£39.79	£34.29	£42.47	£50.13
	£238.74	£205.74	£254.83	£300.80
Rate per metre	£2.39	£2.06	£2.55	£3.01

Example 11.6

Whitewood, sawn purlin/truss member.	m

	150 × 75 mm	100 × 100 mm	200 × 100 mm
Material cost per 100 m	£185.00	£182.00	£370.00
Nails: 1.8 kg at £1.02	£1.84		
2.0 kg		£2.04	
2.3 kg			£2.35
	£186.84	£184.04	£372.35
Waste 10%	£18.68	£18.40	£37.23
Squad laying varying quantities per carpenter hour at £9.30/hour			
150 × 75 mm: 9.5 m	£97.89		
100 × 100 mm: 5.0 m		£186.00	
200 × 100 mm: 2.5 m			£372.00
	£303.41	£388.44	£781.58
Profit and oncost 20%	£60.68	£77.69	£156.32
	£364.10	£466.13	£937.90
Rate per metre	£3.64	£4.66	£9.38

Example 11.7

Whitewood, sawn solid strutting to upper floor or flat roof joists.	m

Joist depth	100 mm	150 mm	175 mm	200 mm
Material all 50 mm thick by depth	£102.00	£127.00	£150.00	£165.00
Nails: 2.80 kg at £0.92/kg	£2.86	£2.86	£2.86	£2.86
	£104.86	£129.86	£152.86	£167.86
Waste 10%	£10.49	£12.99	£15.29	£16.79
Labour: 12.5 hours per 100 m at £9.30	£116.25	£116.25	£116.25	£116.25
	£231.59	£259.09	£284.39	£300.89
Profit and oncost 20%	£46.32	£51.82	£56.88	£60.18
	£277.91	£310.91	£341.27	£361.07
Rate per metre	£2.78	£3.11	£3.41	£3.61

Example 11.8

Whitewood, sawn herringbone strutting to upper floor or flat roof joists with 38 × 32 mm sawn whitewood; assume joists at 450 mm centres.	m

50 mm thick joists at 450 mm centres have 400 mm spaces. There are 100/0.45 = 222 spaces.

A 150 mm joist requires 2 struts approximately 450 mm long; total strutting is 222 × 2 × 0.45 ~ 200 m. For 175 mm joists use 209 m, and for 200 mm joists use 218 m.

Joist depth	150 mm	175 mm	200 mm
Material:			
200 m at £31.25 per 100 m	£62.50		
209 m		£65.31	
218 m			£68.13
Nails: 4.3 kg at £1.02	£4.39	£4.39	£4.39
	£66.89	£69.70	£72.51
Waste 10%	£6.69	£6.97	£7.25
Labour: 20 hours per 100 m of 2 struts at £9.30	£372.00		
		£388.74	
			£405.48
	£445.57	£465.41	£485.24
Profit and oncost 20%	£89.11	£93.08	£97.05
	£534.69	£558.49	£582.29
Rate per metre	£5.35	£5.58	£5.82

Example 11.9

Galvanized steel herringbone strutting; assume joists at 450 mm centres.	m

Steel strutting fits most common centres and joist depths.
There are 222 spaces in 100 m, two struts per space and two nails per strut.

Material: 444 struts at £0.72 each	£319.68
Nails, 38 mm galvanized clout nails, 4 per strut 888 at 515/kg at £1.95/kg	£3.36
	£323.04
Waste 5%	£16.15
	£339.19
Labour: 12 m/hour at £9.30	£77.50
	£416.69
Profit and oncost 20%	£83.34

	£500.03
Rate per metre	£5.00

Example 11.10

145 × 15 mm redwood wrot fascia board and barge board grooved on back for soffit board.	m

Material cost per 100 m	£220.00
Nails: 0.5 kg at £1.32/kg	£0.66
	£220.66
Waste 10%	£22.07
Labour fixing 22 m per hour at £7.16	£42.27
	£285.00
Profit and oncost 20%	£57.00
	£342.00
Rate per metre	£3.42

Example 11.11

32 × 20 mm sawn, Tanalized timber plugged and screwed to masonry.	m

Material cost per 100 m	£18.50
Screws at 1.2 m centres: 83 at £6.50 per 200	£2.71
Plugs: 83 at £2.45 per 100	£2.04
	£23.25
Waste 10%	£2.33
Labour fixing 13 m per hour at £9.30	£71.54
	£97.11
Profit and oncost 20%	£19.42
	£116.54
Rate per metre	£1.17

Example 11.12

9 mm thick WBP plywood in soffit plate, let into groove in fascia/ barge board and pinned to bearer; in width not exceeding 300 mm.	m

From the drawing the estimator sees that the width required is 165 mm. Sheets are supplied 2400 × 1200 mm. There are 1200/165 = 7 strips; 2400 × 45 mm is wasted by the saw or left over. This represents a wastage of 3.75 per cent which is included in the following material cost.

Sheet is 2400 long therefore 7 strips give a total of 16.8 m.

Material cost for 16.8 m strip from 1 sheet	£9.65
Nails: 0.2 kg at £2.30/kg	£0.46
	£10.11
Waste 10%	£1.01
Labour fixing 15 m per hour at £9.30 so cost for 16.8 m	£10.42
	£21.54
Profit and oncost 20%	£4.31
	£25.84
Rate per metre	£1.54

An important departure from previous methods of measurement is the requirement to measure linings by the metre length of wall lined. In stud partition work the partition is measured by the metre length, stating the height (given in stages of 300 mm). The price must include the runners, studs and dwangs plus the plasterboard or other lining on both sides.

The calculation of the number of wall straps, studs or timbers fixed at centres presents a problem. The reason for requiring a fixing centre is that the sheet material which is being fastened to these timbers is of fixed dimension, e.g. gypsum wallboard in 1200 mm wide sheets is most conveniently fixed to timbers at 400 mm centres, i.e. joints between adjacent sheets fall on a timber. How long the wall or wide the ceiling happens to be determines the number of timbers necessary to provide this support.

For example:

An area 8 m wide requires (8000/400) + 1 = 21 = 2.625 studs per metre
An area 6 m wide requires (6000/400) + 1 = 16 = 2.667 studs per metre
An area 4 m wide requires (4000/400) + 1 = 11 = 2.750 studs per metre
An area 3 m wide requires (3000/400) + 1 = 9 = 3.000 studs per metre*
An area 2 m wide requires (2000/400) + 1 = 6 = 3.000 studs per metre
An area 1 m wide requires (1000/400) + 1 = 4 = 4.000 studs per metre*

* In both these instances the dividend is rounded up – no one fixes 'half' a batten.

So it would help to know the approximate length of individual partitions, etc. However, it is impractical to calculate the number of studs or battens for every one so we add a percentage based on an appraisal of the mix of lengths of partitions, etc. In the examples here we have added 10 per cent. So for a metre length of partition with studs at 400 mm centres there are exactly 2.5 studs plus 10 per cent = 2.75 studs × height of the partition.

Example 11.13

Stud partition in height not exceeding 2.40 m, comprising 75 × 38 mm whitewood sawn top and bottom runners, studs at 400 mm centres and three rows of dwangs; 9.5 mm gypsum wallboard both sides. Jointing and decoration by others.	m

Pricing of this item is less prone to error if the logical building sequence is followed

75 × 38 mm sawn softwood in runners

Material cost per 100 m	£60.30	
Nails: 0.8 kg at £1.02	£0.82	
	£61.12	
Waste 10%	£6.11	
	£67.23	
Squad fixing 20 m per hour at £9.30	£46.50	
Cost of 100 m of top or bottom runner	£113.73	
Cost for 2 m of top and bottom runner		£2.27

75 × 38 mm sawn softwood in studding

Material cost per 100 m	£60.30	
Nails: 1 kg at £1.02	£1.02	
	£61.32	
Waste 10%	£6.13	
	£67.45	
Squad fixing 16 m per hour at £9.30	£58.13	
Cost of 100 m of studding	£125.58	

In a 1 m length of partition with studs at 400 centres there will be 2.5 studs + 10% = 2.75 m.

Assume partition full 2.4 m high then there will be (2.4 × 2.75) = 6.6 m of studding

Cost of 6.6 m of studding		£8.29

75 × 38 mm sawn softwood in dwangs (3 in height)

Material cost per 100 m	£60.30	
Nails: 1.4 kg at £1.02	£1.43	
	£61.73	
Waste 10%	£6.17	
	£67.90	
Squad fixing 8 m per hour at £9.30	£116.25	
Cost of 100 m of dwanging	£184.15	

In a 1 m length of partition there will be 3 m of dwangs

Cost of 3 metres of dwangs		£5.52

9.5 mm thick gypsum wallboard nailed to softwood

Material cost per 100 m^2	£143.75	
Nails: 3.75 kg (galvanized wallboard nails) at £2.00	£7.50	
	£151.25	
Waste 10%	£15.13	

	£166.38	
Squad fixing 6.5 m² per hour at £9.30	£143.08	
Cost of 100 m² of plasterboard	£309.45	
Unit length of partition 2.4 m high requires 4.8 m²		
Cost of 4.8 m² of plasterboard		£14.85
		£30.95
Profit and oncost 20%		£6.19
Rate per metre		£37.13

Note that in this example waste is assumed to be a constant 10%. If different percentages were required for the individual components then waste would have to be taken on the cost of the individual materials throughout the calculation.

If the carpenter has no lining material to fix (i.e. where it is fixed by another, such as the plasterer), measurement of battens and open spaced grounds may be expected to follow the traditional route and be measured by the square metre. The following examples show how this work might be priced.

Example 11.14

38 × 19 mm softwood vertical battens at 400 mm centres, plugged to masonry.	m²

Note that the plugging referred to here is achieved by cutting short lengths of timber and driving these into pockets cut into mortar joints. To cover 1 square metre with battens at 400 mm centres requires (1000/400) = 2.5 m material plus ends.

Material cost per 100 m	£17.00
Nails: 0.3 kg at £1.02	£0.31
	£17.31
Timber for plugs 75 mm long at 800 mm centres = 125 plugs; allowing for ends, say, 140 × 75 mm = 10.5 m at £17.00/100 metres	1.78
	£19.09
Waste 10%	£1.91
	£21.00
Squad plugging and fixing 4m per hour at £9.30	£232.50
	£253.50
Add for 'ends' – 10%	£25.35
	£278.85
Profit and oncost 20%	£55.77
	£334.62
Rate per m² (2.5 m of material)	£8.37

Example 11.15

38 × 19 mm sawn softwood branders to ceilings joists, fixed at 400 mm centres.	m^2

Material cost per 100 m	£17.00
Nails: 0.3 kg at £1.02	£0.31
	£17.31
Waste 10%	£1.73
	£19.04
Squad fixing 11 m per hour at £9.30	£84.55
	£103.58
Add for 'ends' – 10%	£10.36
	£113.94
Profit and oncost 20%	£22.79
	£136.73
Rate per m^2 (2.5 m of material)	£3.42

12

Woodwork joinery: second fixings and finishings

SMM7 Sections K11, K20, L10, L20, L30, P20, P21

Introduction

This part of the estimating for woodwork in modern practice is now very much more concerned with the fixing of pre-finished units. The idea of these units is not new. However, the rapid development of new materials and technologies, together with economic pressures on the building industry, have accelerated the demand for and use of pre-finished units. The joiner today is no less skilful than his predecessor, but has new skills and techniques which have boosted his output.

Floors and decks are still boarded with softwood and hardwood strip. However, there are sheet materials now available which one can almost consider to be pre-finished units laid over joists. Indeed, some of these sheet materials have peelable coverings to facilitate leaving a clean floor or deck after the works have been completed.

Doors can now be obtained not just pre-finished but complete in their frames, with architraves fixed one side and loose the other. The frame is placed into the opening in the wall or partition and fastened with frame anchors or screws. The protective polythene wrapper is peeled off, and the door is there complete with all its ironmongery!

Similarly, windows are coming on site already glazed and with all ironmongery factory fitted. They are fixed into openings in walls with framing anchors or brackets and the protective wrapping is removed. The window is then fully functional during building, keeping the weather out and letting light in. Even window boards and bed moulds are part of the kit to be added after fixing the window into position.

The multitude of fasteners available has already been discussed in the previous chapter, as have the various difficulties encountered in working with softwoods and hardwoods.

Before going on to some examples of rates, the reader should remember that we have been working with a squad of 1 labourer and 5 craftsmen costing £9.30 per craftsman hour. This hourly rate will continue to be used in this chapter.

Example 12.1

20 mm thick particle board flooring grade, tongued and grooved all round, nailed to joists with 65 mm improved nails.	m^2

The cover size per sheet is 2400 × 600 mm	
Material cost per 100 m^2	£449.00

Nr of sheets per 100 m^2: $100/(2.4 \times 0.6) = 69.44$
Nails per sheet with joists at 400 mm centres: 28

Nails per 100 m²: 1944 at 229/kg at £27 per 25 kg	£9.17
	£458.17
Waste 5%	£22.91
Labour: craftsman fixes 5 m² per hour at £9.30	
Cost to fix 100 m² is	£186.00
	£667.08
Cost per m²	£6.67
Profit and oncost 20%	£1.33
Rate per m²	£8.00

Example 12.2

An alternative method of calculation for the item in Example 12.1 would be based on the cost per m² for the particle board sheets, as follows.

Material cost per m²	£4.46
Nails per m²: 28 at 229/kg at £27 per 25 kg	£0.13
	£4.59
Waste 5%	£0.23
Labour: craftsman fixes 5 m² per hour at £9.30:	
Cost to fix 1 m²	£1.86
	£6.68
Profit and oncost 20%	£1.34
Rate per m²	£8.02

Example 12.3

19 mm thick softwood tongued and grooved flooring twice nailed to joists; joists at 450 mm centres.	m²

Material cost per 100 m, 150 mm wide	£136.80
coverage is nominally 100.00 × 0.135 = 13.5 m²	
Nails: 65 mm bullet headed brads, 2 per joist, 245*	
joists per 100 m: 490 at 229/kg at £24 per 25 kg	£51.35
	£188.15
Waste 5%	£9.41
Material cost for 15 m²	£123.74
Material cost for 1 m²	£9.17
Labour: craftsman lays 4 m² per hour at £9.30: 1 m²	£2.33

	£11.49
Profit and oncost 20%	£2.30
Rate per m^2	£13.79

* Includes 10% for 'ends'. (See Chapter 11 on timbers at centres.)

Example 12.4

3.2 mm thick standard hardboard pinned to open spaced grounds with 15 mm zinc plated lost head panel pins.	m^2

Material: 2400 × 1200 sheet cost: per m^2	£1.25
Nails: framing at 400 mm centres gives 2.75 m* of nailing each way at 50 mm centres: 168 panel pins at 900/kg at £1.85/kg	£0.35
	£1.60
Waste 5%	£0.08
	£1.68
Labour: craftsman fixes 4 m^2 per hour at £9.30: 1 m^2 Cost to fix 1 m^2	£2.33
	£4.00
Profit and oncost 20%	£0.80
Rate per m^2	£4.80

* Includes 10% for 'ends'. (See Chapter 11 on timbers at centres.)

Example 12.5

Standard casement window in pre-treated softwood to BS 644 Part 1, type 130V, overall size 600 mm wide × 900 mm high, fully weather stripped and fitted into opening in brick wall with proprietary framing fasteners.	Nr

Casement window	£45.00
Fischer fasteners: 4 at £0.56	£2.24
Waste 2.5%	£47.24
Labour: joiner fixes window in 0.5 hours at £9.30	£4.65
	£51.89
Profit and oncost 20%	£10.38
Rate per window	£62.27

Example 12.6

Standard casement window in pre-treated softwood to BS 644 Part 1, type 240TW, overall size 1200 mm wide × 1200 mm high, fully weather stripped and fitted into opening in brick wall with framing fasteners as Fischer's XYZ.	Nr

Casement window	£76.00
Fischer fasteners: 6 at £0.56	£3.36
	£79.36
Waste 2.5%	
Labour: joiner fixes window in 0.5 hours at £9.30	£4.65
	£84.01
Profit and oncost 20%	£16.80
Rate per window	£100.81

Example 12.7

Satin anodized aluminium casement window stay and two pins, screwed to softwood with matching screws.	Nr

Window stay, pins and screws	£4.35
Waste 5%	£0.22
Labour: joiner fixes set in 0.25 hours at £9.30	£2.33
	£6.89
Profit and oncost 20%	£1.38
Rate per set	£8.27

Example 12.8

Satin anodized aluminium casement window fastener, screwed to softwood with matching screws.	Nr

Window fastener and screws	£3.16
Waste 5%	£0.16
Labour: joiner fixes set in 0.35 hours at £9.30	£3.26
	£6.57
Profit and oncost 20%	£1.31
Rate per set	£7.89

Example 12.9

Prefabricated wrot rebated softwood door frame, ex 63 × 88 mm, door size 838 × 1981 mm.	Nr

Frame	£26.00
Frame fasteners: 6 at £0.56	£3.36
	£29.36
Waste 2.5%	£0.73
Labour: joiner fixes frame into brick opening: 0.5 hours at £9.30	£4.65
	£34.74
Profit and oncost 20%	£6.95
Rate per frame	£41.69

An alternative form of frame is made up from rectangular section timber and provided with stops measured separately.

Example 12.10

Wrot softwood door frame, ex 63 × 38 mm, for door size 838 × 1981 mm, plugged to brickwork.	m

Wrot softwood 63 × 38 mm			
Posts	2	1981	3962
Head	1	838	838
Joints	2	25	50
Legs	2	63	126
Horns	2	50	100
			5076

Material is 5.076 metres at £86.00 per 100 m	£4.37
Timber for plugs ex 63 × 38 mm, 6 at 75 mm: incl. with waste	
Nails: if frame double nailed at three points each leg then 12 nails for fixing. 6 nails required for making. 18 nails at 150/kg at £22 per 25 kg	£2.64
	£7.01
Waste 10%	£0.70
Labour: joiner makes up 2.5 m per hour at £9.30	£18.88
Plugging to brickwork: 0.75 hours at £9.30	£6.98
	£33.56
Profit and oncost 20%	£6.71
Cost per frame (compare Example 12.9)	£40.28
Rate per metre	£7.93

Example 12.11

44 mm thick cellular core, flush panelled, external quality door, overall size 838 × 1981 mm, faced with plywood and with hardwood lippings on vertical edges, finished for painting, hung on 1.5 pairs 100 mm brass butt hinges.	Nr

Door	£68.00
Hinges: 1.5 pairs at £7.50 per pair	£11.25
Screws, 40 × 4.5 mm: 8 per hinge at £12 per 200	£1.44
	£80.69
Waste 2.5%	£2.02
Labour: joiner hangs door in 1.5 hours at £9.30	£13.95
	£96.66
Profit and oncost 20%	£19.33
Rate per door	£115.99

Example 12.12

35 mm thick cellular core, flush panelled, internal quality door overall size 838 × 1981 mm, faced with hardboard and with hardwood lippings on vertical edges, hung on 1 pair 100 mm brassed steel butt hinges.	Nr

Door	£32.65
Hinges: 1 pair at £0.95 per pair	£0.95
Screws, brassed 40 × 4.5 mm: 8 per hinge at £4 per 200	£0.32
	£33.92
Waste 2.5%	£0.85
Labour: joiner hangs door in 1.25 hours at £9.30	£11.63
	£46.39
Profit and oncost 20%	£9.28
Rate per door	£55.67

In Examples 12.13 and 12.14 the cost of the ironmongery item fixed is included in the rate. It is more usual to have the supply of ironmongery the subject of a prime cost sum and to give fix-only items in the bill of quantities. This is illustrated in Examples 12.15 and 12.16.

Example 12.13

5 lever mortice lock and satin anodized aluminium lever furniture and escutcheons fixed with matching screws.	Nr

Lock	£17.00
Lever furniture	£13.60
Escutcheon plates and screws	£4.65
	£35.25
Waste 5%	£1.76
Labour: joiner fixes complete in 1.5 hours at £9.30	£13.95
	£50.96
Profit and oncost 20%	£10.19
Rate per set	£61.16

Example 12.14

Satin anodized aluminium postal plate including slotting door.	Nr

Postal plate and fixings	£7.65
Waste 5%	£0.38
	£8.03
Labour: joiner fixes complete in 1.2 hours at £9.3	£11.16
	£19.19
Profit and oncost 20%	£3.84
Rate per plate	£23.03

Example 12.15

Fit and fix only mortice lock, lever furniture and escutcheon plates.	Nr

Material cost included in prime cost sum	nil
Labour: joiner fixes in 1.5 hours at £9.30	£13.95
Profit and oncost 20%	£2.79
Rate per set	£16.74

Example 12.16

Fit and fix only hat and coat hook.	Nr

Material cost included in prime cost sum	nil
Labour: joiner fixes in 0.08 hours at £9.30	£0.74
Profit and oncost 20%	£0.15
Rate per hook	£0.89

Example 12.17

Open tread staircase 914 mm wide, in European redwood complying with BS 585: Part 1, BS 5395; total going 2700 mm; with 237 × 32 mm strings, 258 × 32 mm treads, 110 × 20 mm dummy risers, 66 × 66 × 1350 mm newel posts top and bottom, newel caps, 63 × 44 mm handrail, 32 × 32 mm balusters at 95 mm centres; one string plugged to wall, other string housed both ends to newels, newels checked to floor joists; the whole left clear for varnishing.	Nr

Materials			
Stair	1	£465.00	£465.00
Newel posts:	2	£23.40	£46.80
Newel caps:	2	£5.22	£10.44
String capping:	2	£14.80	£29.60
Handrail:	1	£18.00	£18.00
Balusters:	29	£2.26	£65.54
Framing anchors	8	£0.84	£6.72
Glue: say			£1.30
Nails 50 mm oval brads: say			£0.60
			£644.00
Waste 2.5%			£16.10
Labour			
Joiner to fit and fix prefabricated stair: 3.5 hours at £9.30			£32.55
Fix newels: 2 at 1.35 m at 2 m/hour at £9.30			£12.56
Fix capping to string: 1 at 3.9 m at 4 m per hour at £9.30			£9.07
Fix balusters: 29 at 0.15 hour each at £9.30			£13.49
Fix newel cappings: 2 at 0.05 hour each at £9.30			£0.93
			£728.69
Profit and oncost 20%			£145.74
Rate per staircase			£874.43

Example 12.18

Skirting 100 × 19 mm softwood with two labours, nailed to softwood.	m

Materials per 100 m	£76.50
Nails: 50 mm oval brads, 2 at 400 mm centres: 550* per 100 m, at 440/kg at £38 per 25 kg	£1.90
	£78.40
Waste 5%	£3.92
Labour: joiner fixes 6.25 m per hour at £9.3	£148.80
	£231.12

Profit and oncost 20%	£46.22
	£277.34
Rate per metre	£2.77

* Includes 10% for 'ends'.

Plugging to masonry can be calculated in the same fashion as earlier examples, whether using proprietary plugs or timber plugs.

Example 12.19

Architrave 100 × 19 mm softwood with two labours, nailed to softwood.	m

Materials per 100 m	£76.50
Nails: 50 mm oval brads, 2 at 400 mm centres: 550* per 100 m, at 440/kg at £38 per 25 kg	£1.90
	£78.40
Waste 5%	£3.92
Labour: joiner fixes 8 m per hour at £9.3	£116.25
	£198.57
Profit and oncost 20%	£39.71
	£238.28
Rate per metre	£2.38

Example 12.20

Door stop 42 × 19 mm softwood, nailed to softwood.	m

Materials per 100 m	£42.30
Nails: 40 mm oval brads, 2 at 400 mm centres:	£2.20
550 per 100 m, at 440 kg at £44 per 25 kg	
	£44.50
Waste 5%	£2.23
Labour: joiner flxes 10 m per hour at £9.30	
	£93.00
	£139.73
Profit and oncost 20%	£27.95
	£167.67
Rate per metre	£1.68

13

Structural steelwork and metalwork

SMM7 Sections G10, G12, L11, L21, L31

Introduction

Structural steelwork and metalwork are specialist elements of the building and as such are normally the subject of a subcontract. It would be extremely rare for the contractor's estimator to be involved in pricing other than, for example, an isolated steel section to be used as a lintel, or the fixing in position of a fabricated metal window or metal balustrade.

SMM7 takes account of the specialist nature of these elements of construction in its requirements for dimensioned drawings from which the specialist can measure the work involved. The relevant SMM7 sections are as follows:

G10 Structural steel framing
G12 Isolated structural metal members
L11 Metal windows/rooflights/screens/louvres
L21 Metal doors/shutters/hatches
L31 Metal stairs/walkways/balustrades

This chapter deals with a selection of these elements.

Isolated structural metal members (G12)

Isolated structural metal members are the most common example of steelwork fixed in position by the contractor's own operatives. Proprietary steel lintels over window and door openings will be fixed in position by bricklayers in the course of their work. Other heavier sections may require craneage.

Example 13.1

Isolated structural plain member in universal beam size 356 × 171 mm × 67 kg/m, as lintel over 4 m door opening, and bolting to padstones (measured separately).	Tonnes

Labour/hire				
For hoisting the beam into position allow:	*hrs*	*m*	*rate*	
Labourer	1		£6.50	£6.50
Bricklayer	1		£8.00	£8.00
Hire 750 kg mobile crane and operative	4.5		£18.00	£81.00
Fixing 8 bolts	0.1	8	£8.00	£6.40
				£101.90

	I	*kg/m*	*t*	
Expressed as a rate per tonne	4.3	67	0.29	£353.70
Materials				
Steel beam cost delivered to site				£420.00
Fixing bolts included with brickwork				nil
				£773.70
Profit and oncost 20%				£154.74
Rate per tonne				£928.44

Strictly, proprietary galvanized steel lintels should be measured by weight; however, clause 6.1 of the General Rules in SMM7 allows proprietary items to be measured by their catalogue reference, and this is assumed in the following example.

The squad is assumed to comprise 2 bricklayers and 1 labourer (see Chapter 7), at a cost of £21.50 per hour.

Example 13.2

Lintel type T1, 155 mm high × 2100 mm long.	Nr

	hrs	*m*	*rate*		
Squad fixing	0.18	2.1	£21.50		£8.13
Material cost				£38.65	
Waste 2.5%				£0.97	£39.62
					£47.74
Profit and oncost 20%					£9.55
Rate per lintel					£57.29

Metal windows/rooflights/screens/louvres (L11)

Metal windows are supplied to site complete and ready for fixing to a timber subframe. Aluminium windows are supplied fully finished and often pre-glazed. Example 13.3 assumes that a timber subframe is already in position and has been priced under the timber item L10.

Aluminium windows are delivered to site with tape protecting the finish and with labelling on the glass. The labelling is to inform operatives that the windows are glazed, and helps to prevent, for example, scaffolding tubes being passed through the glazed window.

Pre-glazed units will require careful handling, storing and placing in position, which is reflected in the rate.

Example 13.3

Type XYZ aluminium window. size 888 × 888 mm, with white acrylic finish, factory clear glazed with 11 mm double glazing units, fixed with screws to timber subframe (separately measured) and finished externally with mastic filler.	Nr

	hrs	*nr*	*rate*	
Labour				
Labourer taking delivery, storing, placing in position	0.75		£6.50	£4.88
Joiner fixing (see Chapters 11 and 12 for squad rate)		1	£9.30	£9.30
Bricklayer mastic	0.5		£8.00	£4.00
Labourer removing tape and cleaning glass: 0.75 hours	0.75		£6.50	£4.88
Materials				
Pre-glazed window				£112.00
Damage risk 5%				£5.60
Mastic and screws: say				£5.30
				£145.95
Profit and oncost 20%				£29.19
Rate per window				£175.14

Metal stairs/walkways/balustrades (L31)

In a similar manner to the metal window described above, metal stairs and balustrades are generally measured by number and by reference to a drawing or a manufacturer's catalogue. If the items are supplied for fixing by the contractor, the estimator must liaise with the manufacturer to determine the method of fixing anticipated. It should be noted that mortices in concrete are measured under SMM7 Section E20 or E41.

It is normal for balustrades and the like to be in sections which are capable of being moved and placed in position by hand. However, if a particularly large component, such as a fire escape staircase, requires mechanical lifting, then the estimator should include for this in the item rate.

14
Plumbing installations

SMM7 Sections R10, R11, Y10, Y11, N13

Introduction

This chapter covers standard plumbing installations. Mechanical heating/cooling/refrigeration systems, and ventilation and air conditioning systems, are more specialized forms of mechanical engineering and are generally carried out by specially trained tradesmen. Examples for these systems will not be given. However, the principles of the plumbing examples can be applied to these other systems, bearing in mind the differences in wage costs and the need for specialized tools and plant, especially for handling bulky or heavy items.

The SMM7 sections are: rainwater pipework/gutters; foul water drainage above ground; pipelines; pipeline ancillaries; and sanitary appliances/equipment.

All-in rates for pipe

Pipe is available in a bewildering variety of materials and types. For the plumber, plastics, metal and composition materials are the most common. In these materials, pipes are manufactured to carry hot and cold water; other liquids, including those that are corrosive, flammable, toxic, etc.; gases; foul water, rain water and waste water; and so on. Such pipes share certain features which make it possible to treat them as a single class for measurement, and also allow us to generalize in the explanation of how to determine all-in rates.

Pipe is supplied in lengths or coils, rigid or brittle materials being in lengths and flexible materials being in coils. The most common length is 6 m. However, some pipe is supplied in shorter or longer lengths, especially where the joint is formed as part of the pipe, as in PVC or cast iron soil pipe with spigot and socket joints. Cast iron pipe and gutters are still supplied in imperial lengths of 6 feet. Variety of lengths for pipe with integral sockets (say 1, 2, 3 and 4 m) reduces waste, because every cut produces a length with a joint and another length without a joint!

Coils of 25, 50, 100 or 150 m are common, but some pipe is supplied in 30 m or even 60 m coils. The coil length is dictated as much by what the plumber wants to keep in stock as by old traditional manufacturing processes. For example, it might not have been possible to extrude more than 30 m of a particular pipe in one operation, and therefore it became standard practice to supply in coils of that length. The pipe must inevitably be cut to length. The shorter are the cut lengths, the greater is the labour, the fewer are the joints in the running length and, depending on the rigidity of the pipe and the appliances served, the fewer are the fastenings to hold the pipe in place.

Pipe material varies in its bending ability as follows:

- Bends can be easily made, e.g. in polyethylene and polybutylene water pipe (although of large radius)
- Bends can be made with apparatus, e.g. in half hard copper pipe (BS 2871 Table X) for hot and cold water supply, using springs or a light bending machine

- Bends cannot be made and elbows or bent couplings have to be used, e.g. in cast iron pipe, uPVC soil pipe or hard thin wall copper pipe (BS 2871 Table Z).

The material is always fully specified so that identification is positive.

There is generally a standard method of fixing the pipe in position. Non-standard fixings must be fully specified in the bill item or preamble. The background to which the pipe is fixed is also given. Backgrounds are tabulated in the general rules to SMM7, 8.3(a) to (e).

Joints

Although joints in the running length are deemed to be included, there is a standard method of jointing for each type of pipe in each situation and the cost of labour and materials used for each must be taken into account in the pricing of the running lengths.

Joints occur of course at fittings, i.e. junctions, bends, elbows, valves, etc., and where pipe is connected to sanitary fittings, tankage, etc. In every instance there is a standard method of forming these joints and connections and they are measured separately from the pipe, although some may be included with the apparatus.

Here it may be as well to define the two types of joint that a plumber can make. A *coupling* is designed to join pipe to pipe only. A *connection* is used to join pipe to sanitary fittings, tanks, cylinders, etc. Both tradesmen and suppliers now use either term indiscriminately. The term 'joint' will be used in this text to mean either type.

When making joints the tradesman will use very small (per joint) amounts of general materials. For example, capillary joints on copper pipework require steel wool to clean the pipe, flux to assist in making the solder flow into the whole joint, and fuel to provide heat for the blow torch or clamps. Capillary joints may or may not have an integral solder ring, and if not then solder must be allowed for. Compression joints will require either jointing compound or PTFE tape to ensure a water/gas tight joint. The ring sealed push fit joints on PVC waste pipe require a lubricant. The threaded joints in galvanized steel pipework require hemp yarn and jointing compound or PTFE tape. Joints in spigot and socketed cast iron pipe require gaskin, molten lead and fuel for the furnace. The list goes on and on.

In practically every instance very small amounts of material are used which would be tedious to quantify for every joint. The usual practice is to make a monetary allowance for items like PTFE tape, jointing compound, etc., either in the preliminary works bill or as a small percentage addition at the time of building up the rate. Easily quantifiable and more expensive materials such as solder in end feed joints, or lead and yarn for cast iron soil pipes, are actually quantified and priced when building up the rate for joints of that kind.

SMM7 has recognized that joints are important to the estimator by classifying couplings and connections as having one, two, three or more ends. However, it has made costing difficult for the estimator by lumping together all fittings up to and including 65 mm diameter in one item. (See, for example, rule 2.3.* in Section R on disposal systems.) The difference in cost between a brass/gunmetal 15 mm tee and one of 28 mm is quite significant. More of this is illustrated in the examples which follow.

Special tools

The plumber requires many special tools besides spanners, grips, etc.: for example, blow torches (now generally fuelled with butane or propane), pipe bending machines, pipe threading machines, testing equipment for boiler flues and gas installations and so on. It would not be sensible to attempt to allocate the costs for these tools over each individual item in the bill of quantities. The best solution is to put the cost into the preliminaries bill. Fuel in connection with any of these tools or small plant items might be included here rather than in the rates. The basis for the calculation of the amount of fuel is the general size of the contract and knowledge of fuel consumption on previous contracts.

Labour costs and outputs

In Chapter 2 it was noted that plumbers are paid at a higher rate than other operatives. Thus all-in hourly rates have to be established for this section of work for the various grades of skilled, semi-skilled and unskilled workers. Plumbers are generally classified as technical, advanced and trained; the first two grades carry out various degrees of supervision. The plumber generally operates with a mate, who may be a lower or similar grade of plumber, or even an unskilled worker or an apprentice. Wages for plumbers, apprentices and mates are generally agreed by the Joint Industry Board for the Plumbing and Mechanical Engineering Services Industry.

The first thing to consider is whether or not the labour constants used are for one craft operative working on his own (this is rare), for one craft operative working in a gang, or for a gang. If the constants are for a gang then the composition of the gang must be known, e.g. two plumbers, a plumber and an apprentice, two plumbers and a labourer, and so on.

Next the cost of this labour has to be considered. This is best illustrated with examples. To set the scene it will be assumed that all-in hourly rates for the following operatives have been calculated:

Plumbers: technical	£11.90
advanced	£10.65
trained	£10.05
Apprentices: third year	£5.85

Now, a single plumber is costed out per hour at his particular rate. For two plumbers in a gang, the gang is costed out as the sum of the individual rates. However, if the constants are for one plumber then the average rate of the individual rates is used. For example:

	Gang rate	Individual rate
1 plumber	–	As grade
2 plumbers, 1 advanced, 1 trained	£20.70	£10.35
2 plumbers as above and a 3rd year apprentice	£26.55	£13.28

It will be noticed that in the calculation of the third individual rate, the contribution made by the apprentice to the gang's work is not included; the gang rate is divided by the number of qualified plumbers. This practice of ignoring the apprentice's contribution is common among estimators. Some do include such a contribution; however, it cannot be equivalent to a full craft operative. First of all, apprentices generally do not work as long a day as craft operatives. A third year plumbing apprentice only works 80 per cent of the hours of a plumber. Second, apprentices are allowed time off to attend technical college. Third, they cannot be as productive simply because they are still learning the craft and require guidance. Finally, their training is given by the craft operative(s) and this will slow the operative down. On balance, if apprentices are included in a squad it is probably best to ignore their contribution to production.

Rates used in the examples in this chapter are based on a gang of one advanced and one trained plumber, costing £20.70 per hour total. All constants given in the tables and elsewhere are based on output per craft operative working in a gang and so will be costed at the average rate of £10.35 per hour.

Copper pipework and fittings

Table 14.1 shows ranges of outputs for copper pipework and fittings. The outputs have been abstracted from a variety of sources and reveal the diversity of opinion common among estimators. There has to be an explanation for these differences. The variable attributable to labour factors has been well rehearsed elsewhere in the text. This section deals with those factors attributable to trade practice and to specification.

Variations in pipework installations

Take, for example, a pipe run along a wall in a warehouse or factory. There will be an uninterrupted flow of work. The operatives will snap a chalk line on the background; fasten clips to the background and mount the pipe in the clips; make joints in the running length as they proceed; and finally tighten the clips. The question of what type of joint is made will be discussed later. The only cutting will be at the ends. The plumber can reasonably be expected to erect better than 5 m per hour in these circumstances.

By contrast, take the situation where the pipe is being erected in the toilet blocks in the same warehouse or factory. The first lengths erected during the plumber's roughing stage will present slightly more difficulty due to the restrictions on space. Further time will be required to set out tees to sanitary fittings. The pipe connecting the plumber's roughing to the sanitary fittings will be even more difficult. There will be many made bends which, although measurable and priced separately, will require the plumber to take more time cutting pipe to the correct length. There will also be

Table 14.1 *Copper pipe to BS 2871, Table X, clipped to background including joints in the running length: range of outputs for plumber in a squad in hours per metre of pipe or per joint or labour*

	Pipe diameter		
	15 mm	22 mm	28 mm
Pipe:			
Screwed to timber background	0.19–0.29	0.20–0.30	0.20–0.30
Plugged and screwed to background	0.21–0.40	0.22–0.50	0.24–0.60
Made bends	0.10–0.20	0.15–0.33	0.24–0.60
Capillary joint:			
Elbow/bent coupling	0.18–0.20	0.18–0.32	0.24–0.40
Tee	0.23–0.26	0.23–0.44	0.30–0.40
Straight connection	0.22	0.22	0.28
Straight connection and back nut	0.28	0.28	0.33
Valve	0.20–0.60	0.20–0.70	0.28–0.85
Compression joint:			
Elbow/bent coupling	0.15–0.24	0.16–0.24	0.18–0.32
Tee	0.20–0.33	0.20–0.33	0.30–0.39
Straight connection	0.16	0.16	0.22
Straight connection and back nut	0.20	0.20	0.26
Valve	0.20–0.60	0.22–0.70	0.28–0.85

proportionately more cutting to length than in any other situation. Moreover, there will be work in confined spaces and under and around the sanitary fittings. Horizontal runs of plastic pipework require more supports than the equivalent runs of metal pipework.

Fittings

Although two European standards (EN 1254-1 1998 and EN1254-21998) have been issued, BS 864 is, at the time of writing, still current for capillary and compression fittings suitable for the copper pipe under discussion. Two kinds of capillary fitting are manufactured, as well as two kinds of compression fitting. With both kinds of fitting there are many variations.

The first type of capillary fitting has a solder ring incorporated by the manufacturer. If a flux is used which cleans the pipe, fittings can be pushed onto the pipe and heat applied for a fast joint. It is hard to believe that it takes 12 minutes to form such a joint between two straight lengths of pipe. However, it must be remembered that the plumber is expected to carry out such tasks for 8 hours a day, 5 days a week – and to plan ahead for the next joint!

The second kind of capillary fitting has no solder in it. The pipe is smeared with flux (which again can be a self-cleaning flux), the fitting is pushed onto the pipe and heat is applied. A stick of solder must then be applied to each end of the fitting and allowed to melt and to be drawn into the joint. This is a much more skilful operation; it is easy to put in too little solder and have a joint that is leaky or weak under pressure, or to put in too much solder and allow it to flow into the pipe and cause an obstruction. The latter is obviously expensive in its use of solder – and both are expensive when they cause problems for the contractor which have to be put right! In this instance, the time of 12–15 minutes per joint seems quite fair.

The two types of compression joint are described as type A, non-manipulative, and type B, manipulative. Manipulative means that the ends of the pipes to which the fitting is attached are worked to a special shape which makes the joint mechanically sound and pressure tight. Non-manipulative means that the fitting has been designed to be mechanically sound and pressure tight without the need to do anything to the pipe.

Polybutylene hot and cold water piping and fittings

Systems specified in bills of quantities will generally be required to comply with BS 7291.

Several manufacturers now supply both polybutylene piping suitable for the full range of water services – hot and cold supply and central heating – and in a range of sizes from 10 mm to 28 mm outside diameter, the same as copper pipe. This is important as the pipe can be substituted directly for copper pipe work in renovation works without the need to change the connections at tanks, cylinders and appliances. Early versions of the pipe suffered from the ingress of oxygen through the pipe wall which was detrimental to appliances in certain circumstances, for example central heating systems and especially sealed, pressurized systems. Most manufacturers can now supply a version with an oxygen barrier extruded into the wall of the pipe.

Purpose designed push fit fittings are also made as part of each manufacturer's system. Generally, the wall of the pipe requires support at each fitting and this is given by the insertion of a shouldered liner into the end of the pipe before pushing it home. Various methods have been developed for allowing the fitting to be taken apart, some involve destruction of the grab ring if a new joint is to be made on the pipe; others allow the joint components to be reused for a limited number of times, sometimes using a special tool. Of course, alloy, non-manipulative, compression fittings can be used to join the lengths of pipe and as connections at appliances, tanks and cylinders. Support for the pipe wall becomes even more important when using compression joints. The pipe cannot be formed for use with manipulative fittings and of course capillary fittings cannot be used.

Push fit fittings are all made on the same principle. They incorporate an 'O' ring seal and a 'grab' ring. The 'O' ring seal is fairly obvious. When the pipe is pushed into the fitting it is pushed past the seal first which is squeezed between the pipe and the body of the fitting making a water tight seal. These seals have been used for half a century on soil waste and ventilation piping. With the 'O' ring seal there is no need for jointing tape or compound, indeed the use of compounds could damage the 'O' ring and the use of tape would prevent the pipe being pushed home into the fitting both resulting in leaks. The pipe is next pushed past the 'grab' ring, generally made of stainless steel. The ring fits round the pipe to be joined but has teeth on the internal edge past which the pipe has to be pushed quite firmly. These teeth point towards the body of the fitting and once the pipe is pushed past them they act like barbs and will not allow the pipe to be pulled back out. So strong is the grip of these barbs that bursting pressures on the joints similar to conventional compression and capillary fittings can be achieved.

A word of warning to students and the inexperienced. *Do not* push your finger into one of these joints. It may slide past the grab ring easily enough but you may find yourself with a cut finger or one which is stuck in the fitting and requiring surgery to have the fitting removed.

Various methods are employed to keep the components of the fittings together. Some manufacturers have a threaded nut over the ends of the fitting through which the pipe ends pass before encountering the seals and grab rings. To take the joint apart, the nut is unscrewed and screwed up again to remake the joint. Others use a captive nut with a quick release device on the grab ring actuated with a special tool. All joints incorporate washers and spacers to keep the seal and grab ring in their respective position within the joint. Generally, grab rings and 'O' rings remain intact each time the joint is taken apart but if the pipe end has to be altered in some way the 'O' ring can be reused unless damaged but the grab ring is destroyed and has to be replaced. All these items are sold separately by the manufacturers.

Bends are easily made in the flexible plastic piping but refuse to stay put unless restrained. This is not a problem where the pipe is laced through joists or stud work but if laid loose over, say, a sub-floor, etc. a former has to be used or the pipe will simply take up the largest radius possible and may encroach into space for other services or components. With the use of formers, the bends are easily made without deforming the pipe cross-section down to radii of approximately 4–5 pipe diameters. The formers are comparatively expensive – much more than a made joint in copper pipe so their use is restricted to those situations where the natural self-positioning of the pipe has to be curbed. They generally have holes for screw fastening to a substrate. Elbows or knuckle bends could be used instead but a made bend is preferable as it does not restrict flow through the pipe.

Any type of pipe clip which can be used with copper pipe can be used with polybutylene pipe but those comprising a metal band are best avoided as cyclical thermal movement of the pipe in the clip can result in the metal cutting through the pipe wall! Snap-in plastic clips or plastic saddles are preferred.

The pipe is available in cut lengths generally of 3, 4 and 6 metres and in coils of anything up to 100 metres. Whichever length is supplied, the pipe has the same flexibility and this flexibility is promoted by the manufacturers as of particular advantage in running the pipe in confined spaces or through multiple timbers as in joisted floors or timber partitions. In particular, holes in these timbers do need to be accurately aligned in any way because of the flexibility of the pipe. The push fit joints are also promoted as being a 'time saver' and therefore a cost cutter. This is very true. Wet the end of the pipe with saliva (proprietary lubricant is available – at a cost) and push it into the fitting. Done quicker than it can be explained. Nor are there any ancillary materials used – joint tape or compund, fuel for blow torches, etc. The basic materials are more expensive than the equivalent copper pipe and alloy fittings but the labour required is much reduced.

Table 14.2 gives labour outputs for running pipe in various situations and for the mounting of various standard fittings.

Table 14.2 *Polybutylene pipe and push fit fittings including joints in the running length; range of outputs for plumber in a squad in hours per metre of pipe or per joint or per labour*

	Pipe diameter		
	15 mm	22 mm	28 mm
Pipe:			
Screwed to timber background	0.19–0.29	0.20–0.30	0.20–0.30
Plugged and screwed to background	0.21–0.40	0.22–0.50	0.24–0.60
Made bends	0.07–0.08	0.09–0.11	0.10–0.20
Push fit joint:			
Elbow/bent coupling	0.06–0.07	0.06–0.11	0.08–0.13
Tee	0.08–0.09	0.08–0.15	0.10–0.13
Straight connection	0.07	0.07	0.10
Straight connection and back nut	0.09	0.09	0.11
Valve	0.07–0.20	0.07–0.24	0.09–0.28
Compression joint:			
Straight connection at tank, etc.	0.16	0.16	0.22
Straight connection and back nut at tank, etc.	0.20	0.20	0.26
Valve	0.20–0.60	0.22–0.70	0.28–0.85

Copper pipe of any grade can be used with these push fit fittings for two reasons. First, the outside diameter of both copper and polybutylene pipe is the same and second, the stainless steel grab rings will dig into and grip the soft copper pipe quite securely. These fittings cannot be used with chromium plated pipe or stainless steel pipe as a sure grip cannot be guaranteed on these harder surfaces.

The plastic push fit fittings are generally more expensive than the equivalent brass compression fittings and dearer still than capillary fittings. However, where DZR (dezincification resistant) fittings have to be used, plastic push fit fittings may turn out to be more economical especially when connection times are taken into account.

There does not appear to be any authoritative set of labour constants for plastic pipe with push fit joints. When contacted, several manufacturers were unwilling to give any figures even within a wide range of outputs per fitting, etc. One or two, however, claimed that output was about a third of that for copper pipe and fittings. I would regard this with some caution as fixing pipe is not dependent on how easy it is to joint! The general procedure for fixing pipe has still to be followed and will take exactly the same time whether one is fixing plastic, copper or any other pipe of the same diameter but taking into account the mass and jointing in the running length. So comparing Table 14.2 with Table 14.1 the reader will see that while outputs for fixing pipe remain the same, only those fittings which are wholly push fit have a reduced output.

Examples of rates for plastics hot and cold water pipe and fittings are given later in the chapter.

Example 14.1

15 mm copper pipe, BS 2871:Part 1, Table X, fixed at 1200 mm centres with copper pipe bands screwed to timber, including capillary couplings in the running length.	m

Pipe is generally bought in 6 m lengths; however, the basis for pricing is 100 m.

Materials	
Pipe 100 m	£120.00
Clips: 100/1.2, say 84 at £6.00 per 100	£5.04
Screws brass nr 8 × 40 mm: (100/1.2), say 84 at £3.78 per 100	£3.18
Straight capillary couplings, integral solder ring: (100/6) + 1, say 18 at 24p each	£4.32
	£132.54
Waste 2.5%	£3.31
	£135.85
Labour	
Pipe plumber fixes 5 m per hour: 100/5 at £10.35	£207.00
Couplings: (100/6) + 1, say 18 at 0.22 hours at £10.35	£40.99
	£383.83
Profit and oncost 20%	£76.77
	£460.60
Rate per metre	£4.61

Example 14.2

> 22 mm copper pipe, BS 2871: Part 1, Table X, fixed at 1200 mm centres with copper pipe bands screwed to timber, including capillary couplings in the running length. m

Materials	
Pipe 100 m	£220.00
Clips: 100/1.2, say 84 at £2.60 per 100	£7.56
Screws brass nr 8 × 40 mm: (100/1.2), say 84 at £3.78 per 100	£3.18
Straight capillary couplings, integral solder ring: (100/6) + 1, say 18 at 65p each	£11.70
	£242.44
Waste 2.5%	£6.06
	£248.50
Labour	
Pipe plumber fixes 4 m per hour: 100/4 at £10.35	£258.75
Couplings: (100/6) + 1, say 18 at 0.22 hours at £10.35	£40.99
	£548.23
Profit and oncost 20%	£109.65
	£657.88
Rate per metre	£6.58

Example 14.3

> 22 mm copper pipe, BS 2871:Part 1, Table X, fixed at 1400 mm centres with copper pipe bands screwed to timber, including capillary couplings in the running length. m

Materials

Pipe 100 m	£300.00
Clips: 100/1.4, say 72 at £14.04 per 100	£10.11
Screws brass nr 8 × 40 mm: (100/1.4), say 72 at £3.78 per 100	
Plugs: (100/1.4) say 72 at £0.67 per 100	£0.48
Straight capillary couplings, integral solder ring: (100/6) + 1, say 18 at 33p each	£24.30
	£334.89
Waste 2.5%	£8.37
	£343.26
Labour	
Pipe plumber fixes 3 m per hour: 100/3 at £7.65	£345.00
Couplings: (100/6) + 1, say 18 at 0.28 hours at £7.65	£52.16
	£740.43
Profit and oncost 20%	£148.09
	£888.51
Rate per metre	£8.89

Example 14.4

Joints, valves and small labours on copper pipework

Where pipe size is ≤65 mm, SMM7 requires all sizes of such joints and labours to be included in one item. It follows that the rates given in this example are not bill rates. How the individual rates in the example are aggregated or proportioned to give a bill rate is explained after the example rates.

Extra over given diameter of polybutylene pipe for made bend:

Pipe diameter	15 mm	22 mm	28 mm
Labour: 0.10 hours at £10.35	£1.04		
0.15 hours at £10.35		£1.55	
0.28 hours at £10.35			£2.90
Profit and oncost 20%	£0.21	£0.31	£0.58
Each	£1.24	£1.86	£3.48

Extra over copper piping ≤65 mm diameter for capillary joints with two ends:

Pipe diameter	15 mm	22 mm	28 mm
Material: bent coupling	£0.50	£1.20	£2.20
Waste 2.5%	£0.01	£0.03	£0.06
Labour: 0.18 hours at £10.35	£1.86		
0.22 hours at £10.35		£2.28	
0.30 hours at £10.35			£3.11
	£2.38	£3.51	£5.36
Profit and oncost 20%	£0.48	£0.70	£1.07
Each	£2.85	£4.21	£6.43

Extra over copper piping ≤65 mm diameter for capillary joints with three ends:

Pipe diameter	15 mm	22 mm	28 mm
Material: tee	£0.92	£2.82	£6.12

Waste 2.5%	£0.02	£0.07	£0.15
Labour: 0.25 hours at £10.35	£2.59		
0.30 hours at £10.35		£3.11	
0.35 hours at £10.35			£3.62
	£3.53	£6.00	£9.90
Profit and oncost 20%	£0.71	£1.20	£1.98
Each	£4.24	£7.19	£11.87

Straight connection with backnut, compression ends, type A, for copper:

Pipe diameter	15 mm	22 mm	28 mm
Material: connection	£1.52	£2.60	£7.00
Waste 2.5%	£0.04	£0.07	£0.18
Labour: 0.35 hours at £7.65	£3.62		
0.59 hours at £7.65		£6.11	
1.35 hours at £7.65			£13.97
	£5.18	£8.77	£21.15
Profit and oncost 20%	£1.04	£1.75	£4.23
Each	£6.22	£10.53	£25.38

Screwdown stop valve, compression joints, type A, for copper pipe:

Pipe diameter	15 mm	22 mm	28 mm
Material: valve	£5.40	£10.74	£27.68
Waste 2.5%	£0.14	£0.27	£0.69
Labour: 0.30 hours at £7.65	£3.11		
0.35 hours at £7.65		£3.62	
0.42 hours at £7.65			£4.35
	£8.64	£14.63	£32.72
Profit and oncost 20%	£1.73	£2.93	£6.54
Each	£10.37	£17.56	£39.27

Gate valve, compression joints, type A, for copper pipe:

Pipe diameter	15 mm	22 mm	28 mm
Material: valve	£9.54	£14.95	£34.56
Waste 2.5%	£0.24	£0.37	£0.86
Labour: 0.30 hours at £7.65	£3.11		
0.35 hours at £7.65		£3.62	
0.42 hours at £7.65			£4.35
	£12.88	£18.95	£39.77
Profit and oncost 20%	£2.58	£3.79	£7.95
Each	£15.46	£22.74	£47.73

Where pipe fittings are measured extra over the pipe in which they occur, the cost of the equivalent length of pipe should be deducted. Where these equivalent lengths are very small or the cost is relatively low, this deduction becomes academic. Accordingly, no deduction for this cost has been made in the rates given in Example 14.4.

Note also that when costing out the extra over items for fittings less than 65 mm diameter with two or three ends, there is a significant range of cost from 15 to 28 mm – and 65 mm fittings are much more expensive! The estimator will have to make allowance for this by assessing the ratio of fittings with one, two and three ends in each contract and proportioning out his rates accordingly.

For example, suppose a bill has 100 m of 15 mm pipework and 20 m of 22 mm pipework. There is one bathroom, one shower room and one WC compartment, plus a kitchen and a utility room. The odds are that most of the fittings with one, two and three ends are on 15 mm pipework, probably in the ratio of at least 5:1. Rates from Example 14.4 would be adjusted as follows:

Fittings with two ends:
15 mm: 5 at £2.52 — £12.60
22 mm: 1 at £3.47 — £3.47

£16.07/6 fittings = £2.68

Fittings with three ends:
15 mm: 5 at £4.08 — £20.40
22 mm: 1 at £6.81 — £6.81

£27.21/6 fittings = £4.54

Soil, waste and ventilation piping

Table 14.3 shows ranges of outputs for soil, waste and ventilation piping.

Table 14.3 *Soil, waste and ventilation piping fixed to a solid background, range of outputs per plumber in a squad in hours per metre and jointing sundries*

	Cast iron pipework		uPVC pipework	
	50 mm	100 mm	50 mm	100 mm
Pipe	0.25–0.50	0.35–1.00	0.27–0.33	0.40–0.50
Bend	0.33–0.56	0.75–1.06	0.20–0.39	0.40–0.72
Single branch	0.50–0.76	0.95–1.50	0.33–0.45	0.60–1.00
Offset 150 mm		0.56–0.65	–	0.39–0.40
Offset 300 mm		0.80–1.06	–	0.45–0.72
Lead per joint	0.91–1.00 kg	1.50–1.81 kg	–	–
Yarn per joint	0.10 kg	0.10–0.20 kg	–	–

Example 14.5

100 mm nominal bore cast iron spigot and socket jointed soil piping with staved lead joints and two piece holderbats at 1.8 m centres, fixed to masonry.	m

Materials:		
Pipe in 1.8 m length at £85/m		£152.99
Holderbat, one per length at £5.40		£5.40
Lead per joint: 1.5 kg at £2.02/kg		£3.02
Yarn per joint: 0.15 kg at £2.60/kg		£0.39
		£161.81
Waste 5%		£8.09
Labour		
Method A:		
Fix length of pipe including joint and holderbat: 1.08 hours at £10.35		£11.18
Method B:		
Fix holderbat, hang pipe: 1.8 m at 0.3 hours/m at £10.35	£5.59	
Run joint: 0.5 hours at £10.35	£5.18	
Including labour by method A only		£181.07
Cost per metre		£100.60
Profit and oncost 20%		£20.12
Rate per metre		£120.72

Example 14.6

Extra over 100 mm nominal bore cast iron soil pipe for 92.5° bend.	Nr

'Equivalent lengths' of cast iron pipe for bend calculations are normally available from manufacturers.

Materials	
Bend	£45.60
Joint: lead as pipe	£5.40
yarn as pipe	£3.02
	£54.02
Waste 5%	£2.70
Labour	
Plumber 0.9 hours at £10.35	£9.32
	£66.04
Deduct cost (Example 14.5) of equivalent length of pipe incl. holderbat (joints cost taken with fitting): equivalent length is 0.254 at £158.39	£22.35
	£43.69
Profit and oncost 20%	£8.74
Rate per bend	£52.43

Example 14.7

Extra over 100 mm nominal bore cast iron soil pipe for single branch 92.5°.	Nr

Materials		
Branch		£76.86
Joints 2: lead as pipe, 2 at £1.25		£6.05
yarn as pipe × 2		£0.78
		£83.69
Waste 5%		£4.18
		£87.87
Labour		
Plumber 1.50 hours at £10.35		£15.53
		£103.39
Deduct cost equivalent length pipe: 0.386 at	£158.39	£33.97
		£69.43
Profit and oncost 20%		£13.89
Rate per branch		£83.31

Example 14.8

102 mm bore uPVC soil and ventilating piping with O-ring sealed spigot and socket joints, standard fixings at 2 m centres plugged and screwed to masonry.	m

Materials	
Pipe, 3 m length	£11.30
Standard fixing: 2 at £1.40	£2.80
Screws: £3.80/100, 4 in all	£0.15
Plugs: £2.30/100, 4 at 2.3p	£0.03
	£14.28
Waste 5%	£0.71
Labour	
Plumber 3 m at 0.4 hours/m at £10.35	£12.42
	£27.41
Cost per metre	£9.14
Profit and oncost 20%	£1.83
Rate per metre	£10.96

Example 14.9

Plastics solvent welded boss connection for 32/38 mm waste pipe to uPVC soil pipe.	Nr

Materials	
Connection	£1.56
Waste 2.5%	£0.04
	£1.60
Labour	
Cut hole, solvent weld boss: 0.45 hours at £10.35	£4.66
	£6.26
Profit and oncost 20%	£1.25
Rate per boss	£7.51

Example 14.10

Plastics waste pipe with push fit O-ring seal joints, fixed at given centres with plastics saddles, screwed to timber.	m

Pipe diameter	32 mm	40 mm	50 mm
Materials			
Pipe in 3 m lengths	£1.57	£1.98	£4.93
Saddles 3/1.00 say 3 at 54p.	£1.62	£1.62	£1.62
Screws, say 3 × 2 (30 × 4.5 mm brass) at £3.80/100	£0.23	£0.23	£0.23
Coupling in running length	£0.60	£0.60	£0.94
	£4.02	£4.43	£7.72
Waste 5%	£0.20	£0.22	£0.39
	£4.22	£4.65	£8.10
Labour (Including joints in running length): 3 m at 0.3 hours/m at £10.35 for 32 and 40 pipe	£9.32	£9.32	
3 m at 0.35 hours/m for 50 pipe			£10.87
	£13.53	£13.96	£18.97
Cost per metre	£4.51	£4.65	£6.32
Profit and oncost 20%	£0.90	£0.93	£1.26
Rate per metre	£5.41	£5.59	£7.59

Example 14.11

Extra over plastics waste pipe nominal diameter as given with push fit O-ring sealed joints for fittings.	Nr

Pipe diameter	32 mm	40 mm	50 mm
Material			
Fittings with two ends (bent coupling)	£0.60	£0.60	£0.94
Waste 5%	£0.03	£0.03	£0.05
	£0.63	£0.63	£0.99
Labour Including joints: 0.2 hours at £10.35	£2.07	£2.07	
and 0.30 hours for 50 pipe			£3.11
	£2.70	£2.70	£4.09
Profit and oncost 20%	£0.54	£0.54	£0.82
Rate each	£3.24	£3.24	£4.91
Material			
Fittings with three ends (swept tee)	£0.60	£0.60	£0.94
Waste 5%	£0.03	£0.03	£0.05
	£0.63	£0.63	£0.99
Labour Including joints: 0.33 hours at £10.35	£3.42	£3.42	
and 0.48 hours for 50 mm pipe			£4.97
	£4.05	£4.05	£5.96
Profit and oncost 20%	£0.81	£0.81	£1.19
Rate each	£4.85	£4.85	£7.15

The first thing to note in Example 14.11 is that the cost of the equivalent length of pipe has not been deducted. This length is relatively short and is not considered worth the effort. However, it is on the borderline and some might argue that a deduction should be made.

The second thing is that the rates for fittings would have to be proportioned out in the same way as those for the copper water supply piping in Example 14.4. Also, although the rates for plastics fittings are generally higher, the difference between rates is not as great as in copper capillary fittings, and so proportioning need not have such potentially serious financial consequences.

Example 14.12

22 mm PVC overflow pipe with solvent welded joints, fixed at 600 mm centres with plastics clips screwed to timber	m
and	
Extra over pipe for fittings with ends as given.	Nr

Materials	22 mm pipe	2 ends (bend)	3 ends (tee)	Tank conn.
Pipe in 3 m lengths	£1.07	£0.40	£0.40	£0.40
Clips: 3 at 600 mm centres: say 5 at 8p	£0.40			
30 × 4.5 mm brass screw – 5 at £3.80/100	£0.19			
Joints in running length say every 6 m				
3/6 m = 2 at 30p	£0.80			

	£2.46	£0.40	£0.40	£0.40
Waste 5%	£0.12	£0.02	£0.02	£0.02
Labour (including joints)	£2.58	£0.42	£0.42	£0.42
3 m at 0.22 hours/m at £10.35	£6.83			
0.15 hours at £10.35 for two ended fitting		£1.55	£1.55	
0.20 hours at £10.35 for one end and conn.				£2.07
	£9.41			
Cost per metre/each	£3.14	£1.97	£1.97	£2.49
Profit and oncost 20%	£0.63	£0.39	£0.39	£0.50
Rate per metre/fitting	£3.77	£2.37	£2.37	£2.99

Note that the cost of the equivalent length is once again ignored.

Rainwater goods

Table 14.4 shows ranges of outputs for rainwater goods. Waste is normally 5 per cent, but for excessive short length work allow at least 10 per cent.

Table 14.4 *Rainwater goods, piping fixed to masonry background, gutters to timber background and range of outputs per plumber in a squad in hours per metre and jointing sundries*

	100–120 mm gutter and fittings	65 mm pipe and fittings	100 mm pipe and fittings
*Cast iron goods**			
Straight length	0.30–0.55	0.18–0.54	0.41–0.46
Centre drop	0.25–0.45	–	–
Bend/angle	0.25–0.45	0.18–0.50	0.60–0.81
Stop end	0.12–0.16	–	–
Shoe	–	0.25–0.33	0.38–0.41
uPVC goods			
Straight length	0.26–0.44	0.18–0.54	0.41–0.46
Centre drop	0.25–0.45	–	–
Bend/angle	0.16–0.26	0.15–0.45	0.32–0.48
Stop end	0.10–0.16	–	–
Shoe	–	0.15–0.32	0.24–0.32
Mastic per joint†	0.15–0.20 kg	0.07 kg	0.10 kg

* For heavy gauge cast iron, i.e. 6 mm thick and over, times hours by 1.20 to 1.40.
† This only applies to cast iron goods. Plastics gutters have an integral seal. Down pipes are normally loose jointed; if they have to be sealed, an O or a D ring seal would be part of the pipe as supplied.

Example 14.13

102 mm half round cast iron guttering, BS 640, with bolted joint sealed with mastic, hung and including galvanized steel fascia brackets at 900 mm centres, screwed to timber.	m

Materials	
Gutter in 1.8 m lengths	£25.97
Fascia brackets: 2 at £3.26	£3.26
Screws: 4 at £3.80/100	£0.15
Mastic in joint: 0.25 kg at £2.30/kg	£1.02
M6 gutter bolt and nut: 1 at 48p	£0.48
	£30.88
Waste 5%	£1.54
Labour	£32.42
Fixing brackets, hanging gutter, jointing: 1.8 m at 0.55 hours/m at £10.35	£10.25
	£42.67
Cost per metre	£23.71
Profit and oncost 20%	£4.74
Rate per metre	£28.45

The labour constant of 0.55 hours/m for all work in connection with the gutter is not the only way to tackle this part of the rate. Some estimators prefer to keep constants for three distinct operations, namely fixing a bracket (times two in this instance), handling a length of gutter and making the joint. A very few allocate labour to a fourth category: taking delivery. The latter is a carryover from the days when every piece of material was manhandled off a truck; cast iron rainwater goods took a long time to unload. With palletized goods and mechanical off-loading on site, using either lorry mounted cranes or fork-lift trucks, this is no longer needed. The labour constant for hanging the gutter includes fetching it from storage on site, which is now the only piece of physical carrying done; even this has been reduced on larger sites.

Example 14.14

Extra over 102 mm cast iron gutter for double socketed fittings.	Nr

	Centre drop	Angle	Stop end
Materials			
Fitting	£10.91	£10.52	£4.48
Mastic in joint as gutter:			
2 × 0.25 kg at £2.30/kg	£2.03		
1 × 0.25 kg at £2.30/kg		£1.02	£1.02
	£12.94	£11.54	£5.49
Waste 5%	£0.65	£0.58	£0.27

Labour			
Jointing and fixing: 0.40 hours at £10.35	£4.14	£4.14	
0.14 hours at £10.35			£1.45
	£17.72	£16.26	£7.21
Deduct cost of equivalent length (mm) of gutter at £28.45 centre drop 240, Angle 215. Stop end* 125 mm	£6.83		
		£6.12	
			£3.56
	£24.55	£22.37	£10.77
Profit and oncost 20%	£4.91	£4.47	£2.15
Rate each	£29.46	£26.85	£12.92

* Some patterns of stop end do not displace any gutter and so no equivalent length is deductible.

Example 14.15

> 65 mm cast iron spigot and socket rainwater pipes, loose jointed, with two piece holdfasts to masonry at 1.80 centres. m

Materials	
Pipe in 1.8 m length	£45.43
Holdfast, 2 piece, driven	£4.09
	£49.52
Waste 5%	£2.48
Labour	£52.00
Fixing holdfast, erecting length of pipe: 0.3 hours at £10.35	£5.69
	£57.69
Cost per metre	£32.05
Profit and oncost 20%	£6.41
Rate per metre	£38.46

Example 14.16

> Extra over 65 mm cast iron loose jointed rainwater pipe for fittings. Nr

	Bend	Offset 150 mm	Shoe
Materials			
Fitting	£14.42	£22.16	£14.46
Holdfast	£4.09	£4.09	£4.09

	£18.52	£26.26	£18.55
Waste 5%	£0.93	£1.31	£0.93
Labour	£19.44	£27.57	£19.48
Holdfast and fixing: 0.45 hours at £7.65	£4.66	£4.66	
0.33 hours at £7.65			£3.42
	£24.10	£32.23	£22.90
Deduct cost of equivalent length (mm) of pipe at £38.46 (bend 215, offset 350, shoe 165 mm)	£8.27		
		£13.46	
			£6.35
	£15.83	£18.76	£16.55
Profit and oncost 20%	£3.17	£3.75	£3.31
Rate each	£19.00	£22.52	£19.86

Example 14.17

112 mm uPVC half round gutter, BS 4567, on standard fascia brackets at 900 mm centres, screwed to timber.	m

Materials	
Gutter in 4 m lengths with integral joints	£5.90
Fascia brackets: 4/0.9, say 5 at 46p	£2.30
Screws: 5 × 2 at £3.80/100	£0.38
	£8.58
Waste 5%	£0.43
Labour	
Fixing brackets, hanging gutter, jointing: 4 m at 0.35 hours/m at £10.35	£14.49
	£23.50
Cost per metre	£5.87
Profit and oncost 20%	£1.17
Rate per metre	£7.05

Example 14.18

Extra over 112 mm uPVC gutter for the following fittings (fittings are usually double socketed where appropriate).	Nr

	Centre drop	Angle	Stop end
Materials			
Fitting	£1.57	£1.84	£0.81
Fascia bracket	£0.46	£0.46	
Screws: 2 at 3p	£0.08	£0.08	
	£2.11	£2.38	£0.81
Waste 5%	£0.11	£0.12	£0.04
Labour			
Fixing brackets and jointing: 0.4 hours at £10.35	£4.14	£4.14	
0.1 hours at £10.35			£1.04
	£6.35	£6.63	£1.89
Deduct cost of equivalent length (mm) of gutter at £4.23/m	£1.41	£1.41	–
(centre drop 200, angle 200, stop end 0 mm)			£0.00
	£4.94	£5.22	£1.89
Profit and oncost 20%	£0.99	£1.04	£0.38
Each	£5.93	£6.27	£2.26

Equivalent lengths for PVC rainwater goods are obtained from manufacturers' catalogues or by direct measurement of samples; there are no standard sizes. Stop ends are generally made in two varieties: those fitting over the end of the gutter, and those fitting inside the socket joint of the length of gutter. In either instance there is no saving in the overall length of gutter used.

Example 14.19

68 mm uPVC rainwater pipe, BS 4576, loose jointed with PVC pipe clips at 900 mm centres, plugged and screwed to masonry.	m

Materials	
Length of pipe 3 m	£4.72
Pipe clips: 3/1.00, say 3 at 43p	£1.29
Screws: 3 × 2 at £3.80/100	£0.23
Plugs: 3 × 2 at £0.67/100	£0.04
	£6.28
Waste 5%	£0.31
Labour	
Fixing clips and erecting pipe: 3 m at 0.3 hours/m at £10.35	£9.32
	£15.91
Cost per metre	£5.30
Profit and oncost 20%	£1.06
Rate per metre	£6.36

Example 14.20

Extra over uPVC loose jointed rainwater pipe >65 mm diameter for the following fittings.	Nr

	Bend	Offset 150 mm	Shoe
Materials			
Fitting	£1.20	£3.43	£1.19
Pipe clips: 1 at 43p	£0.43	£0.43	£0.43
Screws: 2 at £3.80/100	£0.08	£0.08	£0.08
Plugs: 2 at £0.67/100	£0.01	£0.01	£0.01
	£1.72	£3.95	£1.71
Waste 5%	£0.09	£0.20	£0.09
Labour			
0.4 hours at £10.35	£4.14	£0.04	
0.3 hours at £10.35			£3.11
	£5.95	£4.19	£4.90
Deduct cost of equivalent length (mm) of pipe at £3.94/m	£1.37		
(bend 215, offset 350, shoe 165)		£2.23	
			£1.05
	£4.58	£1.96	£3.85
Profit and oncost 20%	£0.92	£0.39	£0.77
Each	£5.49	£2.35	£4.62

Plastic water supply piping

Table 14.5 shows ranges of outputs for plastics water pipe.

Table 14.5 *Plastics water pipe: range of outputs per plumber in a squad in hours/m or per fitting. Note not applicable to polybutylene H&C water piping with push fit joints*

	Nominal bore			
	$\frac{1}{2}$ in (12.7 mm)	$\frac{3}{4}$ in (19 mm)	1 in (25.4 mm)	$1\frac{1}{4}$ in (32 mm)
Black polythene type 32 water pipe to BS 1972 for above ground use				
Clipped to timber	0.10–0.33	0.12–0.48	0.14–0.60	0.17–0.70
Clipped to masonry	0.20	0.24	0.28	0.34
Blue polythene water pipe to BS 6572 for underground use				
Laid in trench	0.01–0.06	0.013–0.09	0.015–0.12	0.02–0.15
Any fitting	0.50	0.60	0.70	0.80

Example 14.21

$\frac{1}{2}$in nominal bore Class B polythene piping, to BS 1972, fixed with plastics clips at 600 mm centres, screwed to timber background, including joints in the running length. m

Materials	
Pipe in 50 m coils	£75.00
Clips*: 50/0.6, say 84 at 8.5p	£7.14
Screws: £3.80/100, 1 per clip	£3.19
Joints: take 1 every 25 m on average:	
Type A couplings including liners: say 2 at £4.86	£9.72
	£95.05
Waste 5%	£4.75
Labour	
Plumber 50 m at 0.1 hours/m at £10.35	£51.75
	£151.55
Cost per metre	£3.03
Profit and oncost 20%	£0.61
Rate per metre	£3.64

* Clips are one-piece snap type with single screw fixing.

Example 14.22

Extra over polythene pipe ≤65 mm diameter for type A compression fittings with ends as follows. Nr

	Two ends (reducer)	Three ends (tee)
Materials		
Fitting including liners	£7.30	£8.47
Waste 5%	£0.36	£0.42
Labour		
0.6 hours at £10.35	£6.21	
0.9 hours at £10.35		£9.32
	£13.87	£18.21
Profit and oncost 20%	£2.77	£3.64
Each	£16.64	£21.85

Fittings on polythene pipework are expensive. The estimator faced with the task of proportioning the rates for fittings with two and three ends must exercise considerable care, lest he underprice or overprice the job.

Example 14.23

Full value rate for type A compression fittings to polythene pipework.	Nr

	Connection + backnut	SD valve
Materials		
Fitting including pipe liners	£10.10	£23.29
Waste 5%	£0.51	£1.16
Labour		
Plumber 0.6 hours at £10.35	£6.21	
0.9 hours at £10.35		£9.32
	£16.82	£33.77
Profit and oncost 20%	£3.36	£6.75
Each	£20.18	£40.53

Sanitary fittings

Table 14.6 shows outputs for sanitary fittings. The wide range of outputs in the table seems to depend on whether or not the estimator includes traps, taps, wastes, plugs, chains and chain stays in the overall hours (see the examples which follow).

Table 14.6 *Sanitary fittings: range of outputs per plumber in a squad in hours per unit*

Fitting	Hours
Close coupled WC	1.70–2.35
Low level WC	2.25–4.00
Pedestal wash hand basin	2.05–3.10
Wash hand basin on brackets	1.75–2.50
Rectangular bath: cast iron	3.00–7.00
acrylic	2.65–6.00
Bidet with plain rim	1.95–2.25
Shower slab, control and shower head on riser	6.00
Urinals: stall (per metre)	3.00–5.00
bowl	0.90–2.00
Vitreous china sink	1.25–2.75
Stainless steel sink	1.90–4.50

Example 14.24

White vitreous china washdown WC, BS 1213, with low level cistern, flushing pipe, plastic seat and cover and push fit connection to uPVC soil pipe (timber floor, masonry wall).	Nr

Materials	
WC pan and cistern, washers and fasteners	£125.50
Seat and cover	£28.00
Screws: £7.50/100, 6 at 3p	£0.45
Plugs: £2.30/100, 4 at 2.3p	£0.09
Connector	£2.75
	£156.79
Waste 2.5%	£3.92
Labour:	
Plumber 3.5 hours at £10.35	£36.23
	£196.94
Profit and oncost 20%	£39.39
Rate each	£236.32

The connection for the incoming cold water supply to the cistern should be given in the bill as a separate full value item. So too should the tap connectors for the sanitary fittings in the examples which follow.

Example 14.25

White vitreous enamel pressed steel bath, BS 1390, complete with CP bath mixer taps, CP waste, plastic trap with combined overflow, plug, CP chain and chain stay integral with CP overflow fitting.	Nr

Materials	
Bath	£219.00
Mixer set	£172.00
Waste 38 mm with combined overflow and chain stay	£3.65
Bath trap incl sealing washers at waste	£3.20
	£397.85
Waste 2.5%	£9.95
Labour	
Plumber 3 hours at £10.35	£31.05
	£438.85
Profit and oncost 20%	£87.77
Each	£526.62

Example 14.26

White vitreous china wash hand basin, BS 1188 and BS 3402, complete with white vitreous china pedestal, plastics bottle trap, 32 mm CP waste, plug, chain, chain stay and two 1/2 in. CP taps.	Nr

Materials	
Basin	£38.21
Pedestal	£33.59
Waste fitting incl. plug chain and stay	£3.92
Bottle trap	£5.09
Sealing washers at waste	£0.92
Screws: 4 at £7.22/100	£0.29
CP monobloc tap set	£82.03
	£164.05
Waste 2.5%	£4.10
Labour	
Plumber 3 hours at £7.65	£31.05
	£199.21
Profit and oncost 20%	£39.84
Rate each	£239.05

Water tanks and cylinders

Table 14.7 shows ranges of outputs for water and gas equipment. Central heating systems are dealt with in the next section.

The reader may well find outputs per unit in the region of 5–7 hours per tank or hot water cylinder. Outputs of this magnitude include cutting all holes and fitting overflows and supply pipes in and out of the tank, as well as ball valves, standing wastes, etc. The number of these is obviously an average.

Table 14.7 *Tanks, cylinders and boilers. Range of outputs per plumber in a squad in hours per unit*

Fitting	Hours
Water tanks and cylinders	
PVC tank, BS 4213, loose lid: 4 gal (18 litres)	0.50–0.60
50 gal (227 litres)	1.00–1.25
Direct, indirect and primatic HW cylinders, 30 gal (136 litres)	0.50–0.70
Combination tank/hot water cylinder	0.90–1.00
Cutting holes in tanks and cylinders	0.12
Gas boilers for small domestic CH, balanced flue	
Floor mounted	3.75–4.10
Wall mounted	3.75–4.10
Multipoint boiler	2.50
Electric immersion heater: top entry	0.10–0.15
side entry, incl. cutting hole	1.00–1.25
Radiators up to 2.00 m long and 1.00 m high	
Per hanging bracket*	0.075–0.15
Fitting air bleed valve, blank plug, lock shield valve and control valve	0.50–0.75

* Radiators up to 1 m wide are generally hung on two brackets and up to 2 m wide on three brackets.

Example 14.27

Plastics cold water storage cistern K40R, 213 litres actual capacity, with lid, 32 mm overflow fitting, 22 mm ballcock with plastic ball and tail for plastics pipe and Byelaw 30 kit.	Nr

Materials	
Tank and lid	£53.20
Ballcock	£8.50
Byelaw 30 kit BK40C	£34.00
	£95.70
Waste 2.5 %	£2.39
Labour	
Setting tank: 1 hour at £10.35	£10.35
Cut holes for supply, etc. piping: say 4 × 0.12 hours at £10.35	£5.38
Ballcock: 0.50 hours at £10.35	£5.18
Byelaw 30 kit, 0.5 hours at £10.35	£5.18
	£124.17
Profit and oncost 20%	£24.83
Rate each	£149.01

Generally the connections for joining pipes to tanks, cylinders, boilers, etc. are measured as full value separate items in a bill of quantities, so that only necessary hole cutting is taken with the appliance.

Example 14.28

Indirect copper hot water storage cylinder, BS 1566 Part 1, 136 litres capacity, complete with threaded connection for top entry immersion heater and factory applied foam plastics insulation.	Nr

Materials	
Cylinder	£112.00
Waste 2.5%	£2.80
Labour	
Setting 0.6 hours at £10.35	£0.06
Fitting immersion 0.1 hours at £10.35	£1.04
	£115.89
Profit and oncost 20%	£23.18
Rate each	£139.07

Holes for pipe connections are never cut in an HW cylinder; the manufacturer supplies the cylinder with suitable male or female threaded bosses or even compression or push fit couplings to which connectors or the pipe itself are fastened.

Immersion heater connections are now generally made through a special boss in the top or side of the cylinder. However, it is still possible to have an immersion connected through an oval hole cut in the side of the cylinder; in that case the time taken to fit the immersion heater would be increased to approximately 1 hour. It is assumed that such an immersion heater will be given as a separate item in the bill. On occasions the item for the supply of the immersion heater is given in the electrical bill, and only fitting and attendance are given in the plumbing work.

Central heating systems

Selected outputs are given in Table 14.7.

Example 14.29

Wall mounted gas fired boiler, approximately 8–12 kW output, with a balanced flue.	Nr

Materials	
Boiler and flue	£620.00
Sealing flue to outer wall	£4.65
	£624.65
Waste 2.5%	£15.62
Labour	
Fitting flue through wall, cutting hole and setting boiler*	
4 hours at 10.35	£41.40
	£681.67
Profit and oncost 20%	£136.33
Rate each	£818.00

* Drilling hole through the wall can vary enormously dpending on the wall construction. Thick stone walls with very hard stone can take 6–8 hours alone.

Connections for pipework, pump, controller and external thermostats are usually measured – and priced – separately. The boiler in this example includes a factory fitted and set safety valve and limiting thermostats. Boiler packages are now available where pump and controller are not only supplied but plumbed into the casing. Even a pressure vessel can be supplied plumbed in. In such cases only a cold feed, an open vent and the flow and return pipes are required, plus electrical and gas connections.

Example 14.30

Microbore copper piping, BS 2871 Table Y, including capillary joints in the running length and clips at 500 mm maximum centres nailed to timber or masonry.	m

Pipe diameter	8 mm	10 mm	12 mm
Materials			
Pipe (in 25 m coils), per 100 m	£92.00	£116.00	£140.00
Joints in running length: say every 6 m: say 17 at £1.30	£22.10		

say 17 at 70p		£11.90	
say 17 at 30p			£5.10
Clips: every 500 mm say 200 at £6.00/100	£12.00	£12.00	£12.00
	£126.10	£139.90	£157.10
Waste 5%	£6.31	£7.00	£7.86
Labour			
100 m at 0.1 hours/m at £10.35	£103.50	£103.50	£103.50
	£235.91	£250.40	£268.46
Cost per metre	£2.36	£2.50	£2.68
Profit and oncost 20%	£0.47	£0.50	£0.54
Rate per metre	£2.83	£3.00	£3.22

The inclusion of joints in the running length at 6 m intervals may seem peculiar, as it has been stated that the coils are 25 m long. Why not a joint every 25 m? The reason is that it is unlikely that the plumber will be able to lay more than 6 m at a time on average because of obstructions. The pipe is fully annealed, and theoretically it should be possible to draw the whole 25 m through a series of obstructions. In practice the pipe hardens and can kink and fracture, so drawing in is kept to a minimum. It would also be theoretically possible to apply heat to the pipe and anneal any portions which become work hardened. However, heating costs fuel and time, and this cost is far greater than that of the odd additional joint (three in this case). Heating also carries a fire hazard, and the inner surface of the pipe can become oxidized if the plumber keeps annealing it. Clips for microbore piping are supplied with a hardened steel nail which can be driven direct into masonry as well as timber.

Example 14.31

22 mm brass one-way flow manifold with capillary joints to copper and 4 × 10 mm side entries with type A compression joints to copper.	Nr

Materials	
Manifold	£22.00
Waste 2.5%	£0.55
Labour	
Capillary joint each end 2 × 0.28 hours at £10.35	£5.80
Compression joints 4 × 0.15 hours at £10.35	£6.21
	£34.56
Profit and oncost 20%	£6.91
Rate each	£41.47

Example 14.32

22 mm brass one-way flow manifold with 4 × 8 mm linear flow connections, all capillary joints.	Nr

Materials	
Manifold	£17.50
Waste 2.5 %	£0.44
Labour	
Plumber 0.45 hours at £10.35	£4.66
	£22.60
Profit and oncost 20%	£4.52
Rate each	£27.11

Example 14.33

Stelrad K1 pressed steel pre-finished radiators with factory fitted top and end panels, hung on brackets supplied, each screwed and plugged to masonry; one single entry control valve and one single entry TRV, all with tails for copper pipe. Sizes as given. Nr

		Radiator size	
	600 × 600 high	1000 × 600 high	1200 × 600 high
Materials: Rating W	744	1240	1488
Radiator	£22.00	£36.00	£44.00
1 CP valve at £4.50	£4.60	£4.60	£4.60
1 TRV at £19.60	£19.60	£19.60	£19.60
Screws: 6 at £3.78/100	£0.23	£0.23	
9 at £3.78/100			£0.34
Plugs: 6 at 67p/100	£0.04	£0.04	
9 at 67p/100			£0.06
	£46.47	£60.47	£68.54
Waste 2.5%	£1.16	£1.51	£1.71
Labour:			
Fixing brackets:			
2 × 0.1 hours at £7.65	£2.07	£2.07	
3 × 0.1 hours at £7.65			£3.11
Fitting valves, blanking plug and air cock, hang radiator inc. 2 conns to copper: 0.6 hours at £7.65	£6.21	£6.21	£6.21
	£55.91	£70.26	£79.57
Profit and oncost 20%	£11.18	£14.05	£15.91
Rate each	£67.09	£84.31	£95.48

Example 14.34

50 mm thick glass fibre, polythene clad insulation set for rectangular water storage tank fixed with bands and clips supplied. Nr

Materials	
Insulation set	£36.05
Waste 5 %	£1.80
Labour	
Plumber 0.75 hours at £10.35	£7.76
	£45.62
Profit and oncost 20%	£9.12
Rate each	£54.74

Example 14.35

> Preformed foamed plastic pipe insulation, 19 mm wall thickness, with continuous self fastening edge. m

Materials			
Pipe diameter	15*	22	28
Pipe insulation Byelaw 45 approved*	£5.55	£3.70	£4.50
Waste 5%	£0.28	£0.19	£0.23
Labour			
Plumber fixes 1 m in 0.08 hours @ £10.35/hr	£0.83	£0.83	£0.83
	£6.66	£4.71	£5.55
Profit and oncost 20%	£1.33	£0.94	£1.11
Rate per metre.	£7.99	£5.66	£6.66

* Wall thickness of insulation for 15 mm pipe is 25 mm, other pipe sizes have a wall thickness of 19 mm.

Polybutylene hot and cold water piping and fittings

Example 14.36

> *Polybutylene pipe, BS 7291, fixed at 1000 mm centres with plastic pipe clips, screwed to timber, including push fit straight couplings in the running length. m

Pipe size	15 mm	22 mm	28 mm
Materials			
Pipe in 6 m lengths	£7.65	£14.80	£20.30
Straight coupling in running length	£1.65	£2.25	£5.85
Pipe liners – 2	£0.46	£0.58	£0.70
Bulldog pipe clip – one screw fixing: 600 mm centres	£0.64		
800 mm centres		£0.72	
1000 mm centres			£0.80
30 × 4.5 mm screws at £3.80/100	£0.04	£0.04	£0.04

	£10.44	£18.39	£27.69
Waste 5%	£0.52	£0.92	£1.38
	£10.96	£19.31	£29.07
Labour			
Plumber fixes 1 m in 0.24, 0.25 and 0.25 hrs at £10.35/hr	£14.90		
		£15.53	
			£15.53
Plumber fixes a coupling in 0.06, 0.06 and 0.08 hrs at £10.35/hour	£25.86	£34.83	£44.60
	£0.62	£0.62	£0.83
Cost for 6 metre length	£26.48	£35.45	£45.43
Cost for 1 metre	£4.41	£5.91	£7.57
Profit and oncost 20%	£0.88	£1.18	£1.51
Rate per metre	£5.30	£7.09	£9.09

Example 14.37

Extra over given diameter of polybutylene pipe for made bend:

Pipe diameter	15 mm	22 mm	28 mm
Material: pressed steel former	£1.20	£1.80	£2.30
Screw: 30 × 4.5 to timber £2.60/100	£0.03	£0.03	£0.03
	£1.23	£1.83	£2.33
Waste 5%	£0.06	£0.09	£0.12
Labour: 0.075 hours at £7.65	£0.78		
0.10 hours at £7.65		£1.04	
0.15 hours at £7.65			£1.55
	£2.06	£2.95	£3.99
Profit and oncost 20%	£0.41	£0.59	£0.80
Rate each	£2.48	£3.54	£4.79

Extra over polybutylene piping <65 mm diameter for fitting with two ends:

Pipe diameter	15 mm	22 mm	28 mm
Material: straight coupling coupling	£1.80	£2.25	£5.25
pipe liners – 2	£0.46	£0.58	£0.70
	£2.26	£2.83	£5.95
Waste 2.5%	£0.06	£0.07	£0.15
Labour: 0.065 hours at £10.35	£0.67		
0.085 hours at £10.35		£0.88	
0.15 hours at £10.35			£1.09

	£2.99	£3.78	£7.19
Profit and oncost 20%	£0.60	£0.76	£1.44
Each	£3.59	£4.54	£8.62

Extra over polybutylene piping <65 mm diameter for fitting with two ends:

Pipe diameter	15 mm	22 mm	28 mm
Material: bent coupling	£2.10	£2.90	£6.30
pipe liners – 2	£0.46	£0.58	£0.70
	£2.56	£3.48	£7.00
Waste 2.5%	£0.06	£0.09	£0.18
Labour: 0.065 hours at £10.35	£0.67		
0.085 hours at £10.35		£0.88	
0.1.5 hours at £10.35			£1.09
	£3.30	£4.45	£8.26
Profit and oncost 20%	£0.66	£0.89	£1.65
Each	£3.96	£5.34	£9.91

Extra over polybutylene piping >65 mm diameter for fitting with three ends:

Pipe diameter	15 mm	22 mm	28 mm
Material: tee	£2.10	£3.35	£8.85
pipe liners – 3	£0.69	£0.87	£1.05
	£2.79	£4.22	£9.90
Waste 2.5%	£0.05	£0.08	£0.22
Labour: 0.085 hours at £10.35	£0.88		
0.105 hours at £10.35		£1.09	
0.115 hours at £10.35			£1.19
	£3.72	£5.39	£11.31
Profit and oncost 20%	£0.74	£1.08	£2.26
Each	£4.47	£6.47	£13.57

Straight connection with backnut, compression ends, type A, for copper:

Pipe diameter	15 mm	22 mm	28 mm
Material: connection	£1.52	£2.60	£7.00
pipe liner – 1	£0.23	£0.29	£0.35
	£1.75	£2.89	£7.35
Waste 2.5%	£0.04	£0.07	£0.18
Labour: 0.16 hours at £10.35	£1.66		
0.16 hours at £10.35		£1.66	
0.22 hours at £10.35			£2.28

	£3.45	£4.62	£9.81
Profit and oncost 20%	£0.69	£0.92	£1.96
Each	£4.14	£5.54	£11.77

Screwdown stop valve, compression joints, type A, for copper pipe:

Pipe diameter	15 mm	22 mm	28 mm
Material: valve	£5.40	£10.74	£27.68
pipe liners – 2	£0.46	£0.58	£0.70
	£5.86	£11.32	£28.38
Waste 2.5%	£0.14	£0.27	£0.69
Labour: 0.30 hours at £10.35	£3.11		
0.35 hours at £10.35		£3.62	
0.42 hours at £10.35			£4.35
	£9.10	£15.21	£33.42
Profit and oncost 20%	£1.82	£3.04	£6.68
Rate each	£10.92	£18.25	£40.11

Gate valve, compression joints, type A, for copper pipe:

Pipe diameter	15 mm	22 mm	28 mm
Material: valve	£9.54	£14.95	£34.56
pipe liners – 2	£0.46	£0.58	£0.70
	£10.00	£15.53	£35.26
Waste 2.5%	£0.24	£0.37	£0.86
Labour: 0.30 hours at £10.35*	£3.11		
0.35 hours at £10.35*		£3.62	
0.42 hours at £10.35*			£4.35
	£13.34	£19.53	£40.47
Profit and oncost 20%	£2.67	£3.91	£8.09
Rate each	£16.01	£23.43	£48.57

Plastics stop valve with pushfit joints

Pipe diameter	15 mm	22 mm
Material: valve	£6.35	£9.38
pipe liners – 2	£0.46	£0.58
	£6.81	£9.96
Waste 2.5%	£0.17	£0.25
Labour: 0.07 hours at £10.35	£0.72	
0.11 hours at £10.35		£1.14
	£7.70	£11.35
Profit and oncost 20%	£1.54	£2.27
Rate each	£9.25	£13.62

* Compression fittings for copper pipe fitted into plastic pipelines are subject to the same labour constants as when fitted into copper pipelines.

Joints, valves and small labours on plastics pipework

Where pipe size is ≤65 mm, SMM7 requires all sizes of such joints and labours to be included in one item. It follows that the rates given in Example 14.37 are not bill rates. How the individual rates in the example are aggregated or proportioned to give a bill rate is explained below.

Where pipe fittings are measured extra over the pipe in which they occur, the cost of the equivalent length of pipe should be deducted. Where these equivalent lengths are very small or the cost is relatively low, this deduction becomes academic. Accordingly, no deduction for this cost has been made in the rates given in Example 14.37.

Note also that when costing out the extra over items for fittings less than 65 mm diameter with two or three ends, there is a significant range of cost from 15 to 28 mm. The estimator will have to make allowance for this by assessing the ratio of fittings with one, two and three ends in each contract and proportioning out his rates accordingly.

For example, suppose a bill has 100 m of 15 mm pipework and 20 m of 22 mm pipework. There is one bathroom, one shower room and one WC compartment, plus a kitchen and a utility room. The odds are that most of the fittings with one, two and three ends are on 15 mm pipework, probably in the ratio of at least 5:1. Rates from Example 14.37 would be adjusted as follows:

Fittings with two ends:		
15 mm: 5 at £3.96	£19.80	
22 mm: 1 at £5.34	£5.34	
	£25.14	
Divide by 6 fittings		£4.19
Fittings with three ends:		
15 mm: 5 at £4.47	£22.35	
22 mm: 1 at £6.47	£6.47	
	£28.82	
Divide by 6 fittings		£4.80

15
Linings, partitions and surface finishes

SMM7 Sections K10, K31, K41, M10, M20, M30, M40

Plasterboard

Plasterboard is available in various types for different uses, and in various widths, lengths and thicknesses. The main manufacturer of plasterboard in Britain is British Gypsum Limited. Therefore that company's products are described, as they are those most likely to be encountered on a day–to–day basis.

Gyproc wallboard and plank

Gyproc wallboard is a dry lining plasterboard consisting of an aerated gypsum core encased in a durable paper liner, suitable for application to internal surfaces. The boards have one face of an ivory coloured finish for use where decoration will be applied direct, that is where joints are filled and taped, and the other face of a grey finish which can be coated with plaster.

There are three types of edge profile for differing joint requirements:

- Tapered edge for smooth seamless joints
- Square edge for cover strip jointing or plastering
- Bevelled edge for V jointing.

Wallboards are manufactured in thicknesses of 9.5 mm and 12.7 mm and in widths of 600, 900 and 1200 mm, with the exception of square edge boards, which are produced in widths of 900 and 1200 mm only. The most commonly used lengths of board are 1800 and 2400 mm although boards are produced up to 3500 mm long.

Gyproc plank is a dry lining board manufactured to similar specifications as the wallboard. The main differences are that plank is produced either with both faces in grey or with one grey and one ivory, and that it is only available in 600 mm widths and in one thickness of 19.00 mm.

Gyproc Duplex plasterboards

Gyproc Duplex plasterboards combine the same qualities as ordinary plasterboards with a vapour resistant lining on one face for compliance with current Building Regulations.

Thistle baseboard and Gyproc lath

Thistle baseboard is a lining board for use as a base for gypsum plaster to timber ceiling, partition and wall members. It has an aerated core with grey paper surfaces, and is square on all edges. It is normally supplied in 1200 × 900 mm boards.

Gyproc lath is used in the construction of a suspended ceiling system which incorporates rigid support brackets, enabling the application of plaster bonding and finishing coats.

Gyproc thermal board

Gyproc thermal board is a laminated board composed of a Gyproc wallboard bonded to a layer of expanded polystyrene. This gives a lining board with good thermal insulation. It is manufactured in 25, 32, 40 and 50 mm thicknesses. It has a grey paper surface with square edges for plastering, or an ivory surface with tapered edges for filling and taping ready for decoration. The boards are also available with a vapour resisting membrane at the interface of the polystyrene and the wallboard.

Gyproc Fireline board

Gyproc Fireline board is a gypsum plasterboard with the addition of glass fibre and vermiculite in the core, resulting in improved fire protection properties. The boards can be used as a lining to walls, partitions and ceilings and as a casing to steel beams and columns, increasing the structure's fire resistance.

Boards have one face of ivory coloured finish with tapered or square edges, and one grey face. Fireline boards are manufactured in one thickness of 12.7 mm and in widths of 600, 900 and 1200 mm. The 600 mm board is produced in square edge only. The most commonly used lengths of board are 1800 and 2400 mm, although boards are produced up to 3600 mm long.

Table 15.1 gives a range of labour outputs to fix plasterboards. The hours are based on a squad consisting of 2 tradesmen and 1 labourer.

Table 15.1 *Range of labour outputs for plasterboards*

Description	Application	Thickness (mm)	Squad hours per 100 m^2
Wallboard	Walls	9.5	10–15
		12.7	12–18
	Ceilings	9.5	11–17
		12.7	13–20
Plank	Walls	19.0	15–20
	Ceilings	19.0	17–23
Duplex wallboard	Walls	9.5	10–15
		12.7	12–18
Baseboard	Walls	9.5	10–15
	Ceilings	9.5	11–17
Gyproc lath	Walls	9.5	11–16
		12.7	13–19
	Ceilings	9.5	12–18
		12.7	15–21

Continued

Table 15.1 *Continued*

Description	Application	Thickness (mm)	Squad hours per 100 m^2
Thermalboard	Walls	25.0	14–19
		50.0	17–23
	Ceilings	25.0	16–22
		50.0	19–26
Fireline board	Walls	12.7	13–19
	Ceilings	12.7	15–21
Filling joints and scrim	Walls		6–10
	Ceiling		8–12
Filling, taping joints, slurry coat	Walls		8–12
	Ceiling		10–14

Example 15.1

Linings to walls on timber base; Gyproc wallboard 9.5 mm thick, tapered edge; butt jointed, filled with joint filler, tape and joint finish; spot filling; coat whole surface with slurry coat of joint finish; fixing with galvanized nails; to walls 2.1–2.4 m high. m

Materials

Cost of Gyproc wallboard at £193 per 100 m^2	£193.00
Take delivery and stack 100 m^2: 0.75 hours at £6.50	£4.88
Galvanized plasterboard nails to fix 1200 × 2400 mm board at 150 mm centres:	
To timber members at 600 mm centres: 3 × ((2400/150) + 1) = 51	
To dwangs/noggings at 1200 mm centres: 3 × ((1200/150) + 1 − 3) = 18	
Nr nails per sheet: 51 + 18 = 69	
Nr nails per 100 m^2: 69 × 100/(1.2 × 2.4) = 2395.83	
Cost per 100 m^2: 2396 at 110/kg at £2.50/kg	£54.45
Joint tape to half perimeter of board: 1.2 × 2.4 = 3.6 m	
Length of tape per 100 m^2: 3.6 × 100/(1.2 × 2.4) = 125 m	
Cost of tape per 100 m^2: 125 m at £0.03/m	£3.75
	£256.08
Wastage 5%	£12.80
	£268.88
Joint filler inc. wastage: 22 kg per 100 m^2 at £0.42	£9.24
Joint finish inc. wastage: 33 kg per 100 m^2 at £0.85	£28.05
Slurry coat inc. wastage: 25 kg per 100 m^2 at £0.85	£21.25
Take delivery and stack at 40 nr 25 kg bags per hour:	
100 m^2 requires 80 kg: 0.08 hours at £6.50	£0.52

Labour

Squad of 2 tradesmen and 1 labourer will fix 100 m^2 of wallboard in 12 hours:	
Labour cost per hour (2 × £8.00) + (1 × £6.50) = £22.50	
Cost per 100 m^2: 12 × £22.50	£270.00

Fill joints, tape, finish and slurry coat 100 m^2 will take squad 10 hours:	
Cost per 100 m^2: 10 × £22.50	£225.00
	£822.94
Profit and oncost 20%	£164.59
Cost per 100 m^2	£987.53
Rate per m^2	£9.88

However, in accordance with SMM7 the linings to walls are measured in linear metres, the height is given in stages of 300 mm. The estimator should determine from drawings the specific heights of the walls or partitions that are to be lined. The difference between say 2100 mm and 2400 mm will have a great effect on the competitiveness of the rate and ultimately the profitability, as shown by the following:

Walls 2.1 m high:
  rate per metre = cost per m^2 × wall height = £9.88 × 2.1 = £20.75
Walls 2.4 m high:
  rate per metre = cost per m^2 × wall height = £9.88 × 2.4 = £23.71

Example 15.2

Linings to ceilings on timber base; Gyproc wallboard 9.5 mm thick, tapered edge; butt jointed, filled with joint filler, tape and joint finish; spot filling; coat whole surface with slurry coat of joint finish; fixing with galvanized nails; to ceilings not exceeding 3.5 m above floor level.	m^2

Materials	
Cost of Gyproc wallboard at £195 per 100 m^2	£195.00
Take delivery and stack 100 m^2: 0.8 hours at £6.50	£5.20
Galvanized plasterboard nails to fix 1800 × 900 mm board at 150 mm centres	
To timber members at 450 mm centres: ((1800/150) + 1) = 39	
To dwangs/noggings at 900 mm centres: ((900/150) + 1–3) = 12	
Nr nails per sheet: 39 + 12 = 51	
Nr nails per 100 m^2: 51 × 100/(1.8 × 0.9) = 3148.15	
Cost per 100 m^2: 3148 at 110/kg at £2.50/kg	£71.55
Joint tape to half perimeter of board: 1.8 × 0.9 = 2.7 m	
Length of tape per 100 m^2: 2.7 × 100/(1.8 × 0.9) = 167 m	
Cost of tape per 100 m^2: 167 m at £0.03/m	£5.01
	£276.76
Wastage 5%	£13.84
	£290.60
Joint filler inc. wastage: 26 kg per 100 m^2 at £0.42	£10.92
Joint finish inc. wastage: 38 kg per 100 m^2 at £0.85	£32.30
Slurry coat inc. wastage: 25 kg per 100 m^2 at £0.85	£21.25
Take delivery and stack at 40 nr 25 kg bags per hour:	
100 m^2 requires 89 kg: 0.09 hours at £6.50	£0.59

Labour

Squad of 2 tradesmen and 1 labourer will fix 100 m^2 of wallboard in 14 hours:	
Labour cost per hour (2 × £8.00) + (1 × £6.50) = £22.50	
Cost per 100 m^2: 14 × £22.50	£315.00
Fill joints, tape, finish and slurry coat 100 m^2 will take squad 12 hours:	
Cost per 100 m^2: 12 × £22.50	£270.00
	£940.65
Profit and oncost 20%	£188.13
Cost per 100 m^2	£1128.78
Rate per m^2	£11.29

Example 15.3

> Linings to walls on timber base; thistle baseboard 9.5 mm thick, butt jointed, filled with joint filler and thistle board finish; reinforced with jute scrim; fixing with galvanized nails; to walls 2.1–2.4 m high. m

Materials

Cost of thistle baseboard at £195 per 100 m^2	£195.00
Take delivery and stack 100 m^2: 0.9 hours at £6.50	£5.85
Galvanized plasterboard nails to fix 1200 × 900 mm board at 150 mm centres	
To timber members at 450 mm centres: 3 × ((1200/150) + 1) = 27	
To dwangs/noggings at 600 mm centres: 3 × ((900/150) + 1 − 3) = 8	
Nr nails per sheet: 27 + 8 = 35	
Nr nails per 100 m^2: 46 × 100/(1.8 × 0.9) = 2160.49	
Cost per 100 m^2: 2160 at 110/kg at £2.50/kg	£49.09
Joint scrim to half perimeter of board: 1.2 × 0.9 = 2.1 m	
Length of tape per 100 m^2: 2.1 × 100/(1.2 × 0.9) = 194 m	
Cost of tape per 100 m^2: 194 m at £0.09/m	£17.46
	£267.40
Wastage 2.5%	£6.69
	£274.09
Carlite bonding coat filling joints inc. wastage: 26 kg per 100 m^2 at £0.42	£10.92
Take delivery and stack at 40 nr 25 kg bags per hour:	
100 m^2 requires 26 kg 0.03 hours at £6.50	£0.20

Labour

Squad of 2 tradesmen and 1 labourer will fix 100 m^2 of wallboard in 12 hours:	
Labour cost per hour (2 × £8.00) + (1 × £6.50) = £22.50	
Cost per 100 m^2: 12 × £22.50	£270.00
Fill joints and scrim 100 m^2 will take squad 6 hours:	
Cost per 100 m^2: 6 × £22.50	£135.00
	£690.21
Profit and oncost 20%	£138.04

Cost per 100 m^2	£828.25
Cost per m^2	£8.28
Rate per m – 2.4 m high £8.25 × 2.4	£19.87

Cement/sand screeds

Table 15.2 shows a range of labour outputs for laying cement/sand floor screeds. The hours are again based on a squad consisting of 2 tradesmen and 1 labourer.

Example 15.4

Mortar; cement and sand, 1:3; 50 mm work to floors; level and to falls only <15° from horizontal; wood float finish.	m^2

Materials	
Washed sand per tonne	£19.00
Cement bagged, delivered to site per tonne	£90.00
Unload and stack: 1 hour/tonne at £6.50	£6.50
	£96.50
Materials cost per m^3 = tonnes/m^3 at cost/tonne:	
Cement: 1.28 at £96.50	£123.52
Sand: 1.50 at £19.00	£28.50
Mix by volume:	
1 m^3 cement costs	£123.52
3 m^3 sand costs	£85.50
	£209.02
Allow 20% shrinkage: therefore add 25%	£52.26
Cost 4 m^3 of 1:3 mix	£261.28
Cost 1 m^3 of 1:3 mix	£65.32
Cost per m^2 at 50 mm thick: £65.32 × 0.05	£3.27
Wastage 5%	£0.16
Labour	
Squad of 2 tradesmen and 1 labourer will lay 1 m^2 in 0.2 hours:	
Labour cost per hour (2 × £8.00) + (1 × £6.50) = £22.50	
Cost per m^2: 0.2 × £22.50	£4.50
	£7.93
Profit and oncost 20%	£1.59
Rate per m^2	£9.52

Table 15.2 *Range of labour outputs for cement/sand floor screeds*

Description	Squad output
Cement and sand screed wood float:	
25 mm thick	0.16–0.20 hours/m^2
50 mm thick	0.18–0.22 hours/m^2
75 mm thick	0.22–0.26 hours/m^2
Laying to slopes over 15° from horizontal per 25 mm thickness	0.05–0.10 hours/m^2
Skirting 150 mm high with radius cove at floor	0.23–0.27 hours/m

Table 15.3 *Range of labour outputs for granolithic floor screeds*

Description	Squad output
Cement and granite screed steel trowel:	
25 mm thick	0.25–0.29 hours/m^2
50 mm thick	0.30–0.34 hours/m^2
75 mm thick	0.35–0.40 hours/m^2
Laying to slopes over 15° from horizontal per 25 mm thickness	0.05–0.10 hours/m^2
Skirting 150 mm high with radius cove at floor	0.28–0.33 hours/m

Granolithic screeds

Table 15.3 shows a range of labour outputs for laying granolithic floor screeds. The hours are again based on a squad consisting of 2 tradesmen and 1 labourer.

Example 15.5

Granolithic; cement and granolithic, 2:5; 25 mm work to floors; level and to falls only <15° from horizontal; steel trowel finish.	m^2

Materials

Granite chippings per tonne delivered to site in bulk	£21.00
Cement bagged, delivered to site per tonne	£90.00
Unload and stack: 1 hour/tonne at £6.50	£6.50
	£96.50
Materials cost per m^3 = tonnes/m^3 at cost/tonne	
Cement: 1.28 at £96.50	£123.52
Granite chippings: 1.60 at £21.00	£33.60

Mix by volume:

$2\,m^3$ cement costs	£247.04
$5\,m^3$ sand costs	£168.00
	£415.04
Allow 20% shrinkage: therefore add 25%	£103.76
Cost $7\,m^3$ of 2:5 mix	£518.80
Cost $1\,m^3$ of 2:5 mix	£74.11
Cost per m^2 at 25 mm thick: £74.11 × 0.025	£1.85
Wastage 5%	£0.09

Labour

Squad of 2 tradesmen and 1 labourer will lay $1\,m^2$ in 0.27 hours:

Labour cost per hour (2 × £8.00) + (1 × £6.50) = £22.50	
Cost per m^2: 0.27 × £22.50	£6.08
	£8.02
Profit and oncost 20%	£1.60
Rate per m^2	£9.62

Example 15.6

> Granolithic; cement and granolithic, 2:5; 25 mm wide, 150 mm high skirting with 20 mm radius cove next to floor; on brickwork or blockwork base. m

Materials

Cost $1\,m^3$ of 2:5 mix (Example 15.5): £74.11	
Cost 1 m skirting 150 mm high × 25 mm wide: £74.11 × 0.025 × 0.15	£0.28
Cost to form cove: say	£0.02
Cost for temporary grounds: say	£0.03
	£0.33
Wastage 5%	£0.02

Labour

Squad of 2 tradesmen and 1 labourer will produce 1 m in 0.29 hours, including fixing and removing temporary grounds:

Labour cost per hour (2 × £8.00) + (1 × £6.50) = £22.50	
Cost per m: 0.29 × £22.50	£6.53
	£7.20
Profit and oncost 20%	£1.44
Rate per m	£8.64

Table 15.4 *Range of labour outputs for plaster work*

Description	Squad hours per 100 m^2
Float and finish coat on:	
Plasterboard	16–20
Brick or blockwork	18–22
Concrete	16–20
Expanded metal lath	20–24
Two-coat render to brick or blockwork	18–22
Add to above labour outputs for work to:	
Ceilings	15%
Heights above 3.5 m per 1.5 m stage	5%
Widths not exceeding 300 mm	25%
Rounded angles: tradesman m/hour	15–20

Plastered and rendered coatings

Table 15.4 shows a range of labour outputs for applying plaster to walls. The hours are again based on a squad consisting of 2 tradesmen and 1 labourer.

Example 15.7

Plaster to walls; Carlite; pre-mixed; floating coat of bonding 8 mm thick; finishing coat of finish 2 mm thick; 10 mm work to 9.5 mm plasterboard base to walls; steel trowelled; >300 mm wide.	m^2

Materials	
Coverage of Carlite bonding 8 mm thick is 150 m^2 per 1000 kg	
Cost of Carlite bonding per 100 m^2: 100.00/150 at £150.00 per tonne	£100.00
Coverage of Carlite bonding 2 mm thick is 450 m^2 per 1000 kg	
Cost of Carlite bonding per 100 m^2: 100.00/450 at £105.00 per tonne	£23.00
Take delivery and stack at 40 nr 25 kg bags per hour:	
100 m^2 requires 36 bags: 0.90 hours at £6.50	£5.85
	£128.85
Wastage 2.5%	£3.22
	£132.07
Labour	
Squad of 2 tradesmen and 1 labourer will mix and apply 100 m^2 in 18 hours:	
Labour cost per hour $(2 \times £8.00) + (1 \times £6.50) = £22.50$	
Cost per 100 m^2: $18 \times £22.50$	£405.00

	£537.07
Profit and oncost 20%	£107.41
Cost per 100 m^2	£644.49
Rate per m^2	£6.44

Example 15.8

> Plaster to walls; Carlite; pre-mixed; floating coat of browning 11 mm thick; finishing coat of finish 2 mm thick; 13 mm work to walls on brickwork base; steel trowelled; >300 mm wide. m^2

Materials

Coverage of Carlite browning 11 mm thick is 140 m^2 per 1000 kg	
Cost of Carlite browning per 100 m^2: 100.00/140 at £150.00 per tonne	£107.00
Coverage of Carlite finish 2 mm thick is 450 m^2 per 1000 kg	
Cost of Carlite finish per 100 m^2: 100.00/450 at £105.00 per tonne	£23.00
Take delivery and stack at 40 nr 25 kg bags per hour:	
100 m^2 requires 37 bags: 0.93 hours at £6.50	£6.05
	£136.05
Wastage 5%	£6.80

Labour

Squad of 2 tradesmen and 1 labourer will mix and apply 100 m^2 in 20 hours:	
Labour cost per hour (2 × £8.00) + (1 × £6.50) = £22.50	
Cost per 100 m^2: 20 × £22.50	£450.00
	£621.89
Profit and oncost 20%	£124.38
Cost per 100 m^2	£746.27
Rate per m^2	£7.46

Example 15.9

> Plaster to isolated columns; Carlite; pre-mixed; floating coat of bonding 8 mm thick; finishing coat of finish 2 mm thick; 10 mm work to isolated columns on concrete base; steel trowelled; >300 mm wide. m

Materials

Cost 100 m^2 materials inc. wastage (Example 15.7):	£132.07

Labour

Squad of 2 tradesmen and 1 labourer will mix and apply 100 m^2 in 20 hours:	
Labour cost per hour (2 × £8.00) + (1 × £6.50) = £22.50	
Cost per 100 m^2: 20 × £22.50	£450.00

Additional labour for width <300 mm: say 25%	£112.50
Cost per 100 m^2	£694.57
Cost per m^2	£6.95
Assume 150 mm wide columns	
Cost per metre for width 150 mm: £6.95 × (150/1000)	£1.04
Profit and oncost 20%	£0.21
Rate per m	£1.25

Example 15.10

> Render; cement and sand 1:3; first coat 10 mm thick; finishing coat 3 mm thick; 13 mm work to walls on brickwork or blockwork base; steel trowelled; <300 mm wide. m

Materials	
Cost 1 m^2 of 1:3 mix (Example 15.4): £65.32	
Cost per m^2 at 13 mm thick: £65.32 × 0.013	£0.85
Wastage 5%	£0.04
Labour	
Squad of 2 tradesmen and 1 labourer will mix and apply 1 m^2 in 0.2 hours:	
Labour cost per hour (2 × £8.00) + (1 × £6.50) = £22.50	
Cost per 100 m^2: 0.2 × £22.50	£4.50
Additional labour for width <300 mm: say 25%	£1.13
Cost per m^2	£6.52
Assume 150 mm wide	
Cost per metre for width 150 mm: £6.52 × (150/1000)	£0.98
Profit and oncost 20%	£0.20
Rate per m	£1.18

Example 15.11

> Rounded angles and intersections; on cement, sand plaster; 10–100 mm radius. m

Labour	
Tradesman will form 17 m in 1 hour	
Cost per metre: (1/17) 0.06 × £8.00	£0.47
Additional labour for width <300 mm: say 25%	£0.12
Rate per m	£0.59

Table 15.5 *Range of labour outputs for roughcast work*

Description	Squad hours per 100 m^2
Roughcast dry dash	19–24
Roughcast wet dash	22–27

Roughcast coatings

Table 15.5 shows a range of labour outputs for applying roughcast coatings. The hours are again based on a squad consisting of 2 tradesmen and 1 labourer.

Example 15.12

Roughcast: first coat of cement, lime and sand 2:1:9; second coat of cement, lime and coloured sand 1:1:6; third coat of spar chips, dry dash; 19 mm work to walls on brickwork or blockwork base; >300 mm.	m^2

Materials		
Lime bagged delivered to site per tonne	£130.00	
Take delivery and stack: 1 hour/tonne at £6.50	£6.50	
Slaking limc: 2 hours at £6.50	£13.00	
	£149.50	
Materials cost per m^3 = tonnes/m^3 at cost/tonne:		
Cement: 1.28 at £96.50	£123.52	
Lime: 1.67 at £149.50	£249.67	
Sand: 1.28 at £19.00	£24.32	
Coloured sand: 1.28 at £25.00	£32.00	
Chips: 1.60 at £41.00	£65.60	
First coat 2:1:9 mix by volume:		
Cement 2 m^3 at £123.52	£247.04	
Lime 1 m^3 at £249.67	£249.67	
Sand 9 m^3 at £24.32	£218.88	
12 m^3	£715.59	
Allow 20% shrinkage: therefore add 25%	£178.90	
Cost 12 m^3 of 2:1:9 mix	£894.49	
Cost 1 m^3 of 2:1:9 mix	£74.54	
Cost per m^2 at 10 mm thick: £74.54 × 0.019		£1.42
Wastage 5%		£0.07

Second coat 1:1:6 mix by volume:			
Cement	1 m^3 at £123.52	£123.52	
Lime	1 m^3 at £249.67	£249.67	
Coloured sand	6 m^3 at £32.00	£192.00	
	8 m^3	£565.19	
Allow 20% shrinkage: therefore add 25%		£141.30	
Cost 8 m^3 of 1:1:6 mix		£706.49	
Cost 1 m^3 of 1:1:6 mix		£88.31	
Cost per m^2 at 9 mm thick: £88.31 × 0.009			£0.79
Wastage 2.5%			£0.02
Third coat: 1 m^3 chips will cover 110 m^2			
Cost per m^2: £65.60/110			£0.60
Wastage 5%			£0.03

Labour

Squad of 2 tradesmen and 1 labourer will produce 1 m^2 in 0.23 hours, including fixing and removing temporary grounds:

Labour cost per hour (2 × £8.00) + (1 × £6.50) = £22.50	
Cost per 100 m^2: 0.23 × £22.50	£5.18

Mixer

In this example we shall include the cost of a mixer rather than presume that its cost is covered in the preliminaries. The costs for mixing and transporting are included in the above labour costs. A mixer would normally be shared between a number of squads: in this example say 4 squads to share.

Cost of mixer per hour: say £2.50	
Cost per m^2: (0.23 hours at £2.50)/4	£0.14
	£8.25
Profit and oncost 20%	£1.65
Rate per m^2	£9.90

Example 15.13

Roughcast: first coat of cement, lime and sand 2:1:9; second coat of cement, lime and coloured sand 1:1:6; third coat of cement and granite chips 1:2 wet dash; 27 mm work to walls on brickwork or blockwork base; >300 mm.	m^2

Materials

Materials cost per m^3 = tonnes/m^3 at cost/tonne:

Cement: 1.28 at £96.50	£123.52	
Granite chips: 1.60 at £21.00	£33.60	
First coat: cost per m^2 at 10 mm thick (Example 15.12) inc. wastage		£1.49

Second coat: cost per m^2 at 9 mm thick (Example 15.12) inc. wastage		£0.81
Third coat 1:2 mix by volume:		
Cement 1 m^3 at £123.52	£123.52	
Granite chips 2 m^3 at £33.60	£67.20	
3 m^3	£190.72	
Allow 20% shrinkage: therefore add 25%	£47.68	
Cost 3 m^3 of 1:2 mix	£238.40	
Cost 1 m^3 of 1:2 mix	£79.47	
Cost per m^2 at 8 mm thick: £79.47 × 0.008		£0.64
Wastage 5%		£0.03
Labour		
Squad of 2 tradesmen and 1 labourer will produce 1 m^2 in 0.27 hours, including fixing and removing temporary grounds:		
Labour cost per hour (2 × £8.00) + (1 × £6.50) = £22.50		
Cost per 100 m^2: 0.27 × £22.50		£6.08
Mixer		
Cost per m^2: (Example 15.12): (0.27 hours at £2.50)/4		£0.17
		£9.21
Profit and oncost 20%		£1.84
Rate per m^2		£11.06

Beads

Table 15.6 shows a range of labour outputs for fixing beads.

Table 15.6 *Range of labour outputs for beads*

Description	Tradesman m/hour
Metal angle bead	18–20
Metal stop bead	18–20

Example 15.14

Galvanized steel angle beads: fixing with plaster dabs; reference 553.	m

Materials	
Angle bead in 3 m lengths at £1.44: per metre	£0.48
Plaster dabs: per metre say	£0.25
	£0.73
Wastage 5%	£0.04

Labour
Tradesmen will fix 20 m in 1 hour:
Cost per metre: 0.05 hours at £8.00 | £0.40

	£1.17
Profit and oncost 20%	£0.23
Rate per m	£1.40

Metal mesh lathing

Table 15.7 shows a range of labour outputs for fixing expanded metal lath. The hours are again based on a squad consisting of 2 tradesmen and 1 labourer.

Table 15.7 *Range of labour outputs for expanded metal lath*

Description	Squad hours per 100 m²
Expanded metal lath to walls	18–22
Expanded metal lath to ceilings	23–26

Example 15.15

Lathing; galvanized expanded metal; butt joints; fixing with galvanized nails; 6 mm mesh to walls on timber base; >300 mm wide.	m^2

Materials

Cost of steel lath at £314.00 per 100 m^2	£314.00
Take delivery and stack: 1 hour/tonne at £6.50	£6.50

Galvanized nails to fix 2500 × 700 mm sheet at 150 mm centres (sheet overlap, effective area 2400 × 600 mm):
To members at 600 mm centres: 2 × [(2400/150) + 1] = 34
To dwangs at 1200 mm centres: 3 × [(600/150 + 1 − 2] = 9
Nr of nails per sheet: 34 + 9 = 43
Nr of nails per 100 m^2: 43 × 100/(2.4 × 0.6) = 2986

Cost per 100 m^2: at 2986 at 110/kg at £2.50	£67.86
	£388.36
Wastage 5%	£19.42

Labour
Squad of 2 tradesmen and 1 labourer will fix 100 m^2 in 20 hours:

Cost per 100 m^2: 20.00 hours at £22.50	£450.00

	£857.78
Profit and oncost 20%	£171.56
Cost per 100 m^2	£1 029.34
Rate per m^2	£10.29

Fibrous plasterwork

A typical output for a tradesman fixing plaster cove is 8–12 m/hour.

Example 15.16

Fibrous plaster cove; plain; 100 mm girth.	m

Materials	
Preformed cove in 3 m lengths at £2.40: per metre	£0.80
Cove adhesive 5 kg pack at £2.65 will fix 20 m	£0.13
	£0.93
Wastage 5%	£0.05
Labour	
Tradesmen will fix 10 m in 1 hour:	
Cost per metre: 0.10 hours at £8.00	£0.80
	£1.78
Profit and oncost 20%	£0.36
Rate per m	£2.13

Ceramic and quarry tiling

Table 15.8 shows a range of labour outputs for tiling. The hours are again based on a squad consisting of 2 tradesmen and 1 labourer.

Example 15.17

Ceramic tiles; plain white glazed; butt joints; fixing with adhesive; grouting horizontal and vertical joints with anti-fungal cement; 150 × 150 × 5.5 mm units to walls on plaster base; >300 mm wide.	m^2

Materials	
Tiles supplied in packs of 18 tiles at £6.30 per pack	
Area covered by 100 tiles: 100 × 0.152 × 0.152 = 2.31 m^2 (2 mm joints)	
Cost of 100 tiles: (100/18) × £6.30	£35.00
Unload and stack 100 tiles (5.55 packs) in 6 minutes: 0.1 × £6.50	£0.65

Adhesive: 0.6 litres at £0.98 per litre	£0.59
Cement grout: 0.5 kg at £0.83 per kg	£0.42
	£36.65
Wastage 5%	£1.83

Labour
Squad of 2 tradesmen and 1 labourer will fix 100 tiles in 1.10 hours:
Labour cost per hour (2 × £8.00) + (1 × £6.50) = £22.50

Cost per 100 tiles: 1.10 × £22.50	£24.75
	£63.24
Profit and oncost 20%	£12.65
Cost per 100 tiles	£75.88
Rate per m^2: £75.88/2.31 m^2	£32.84

Table 15.8 *Range of labour outputs for tiling*

Description	Squad hours per 100 m^2
Ceramic wall tiles 5.5 mm thick:	
150 × 150 mm	0.38–0.43
100 × 100 mm	0.35–0.40
Quarry tiles to floors 16 mm thick:	
225 × 225 mm	1.35–1.45
150 × 150 mm	1.25–1.35
150 × 150 mm cove skirting	1.75–1.85

Example 15.18

Quarry tiles; butt joints; including 100 mm bed of cement, sand mortar 1:3; jointing and pointing with cement, sand mortar 1:3; to floors level or to falls less than or equal to 15°; 150 × 150 × 16 mm units; plain; symmetrical pattern; >300 mm wide. m^2

Materials
Tiles supplied in packs of 10 tiles at £7.20 per pack
Area covered by 100 tiles: 100 × 0.154 × 0.154 = 2.37 m^2 (4 mm joints)

Cost of 100 tiles: (100/10) × £7.20	£72.00
Unload and stack 100 tiles (10 packs) in 15 minutes: 0.25 × £6.50	£1.63
Mortar bed for 100 tiles at 10 mm thick: 0.0237 m^3 at £65.32/m^3	£1.55
Mortar grout: say 0.005 m^3 at £65.32/m^3	£0.33
	£75.50
Wastage 5%	£3.78

Labour

Squad of 2 tradesmen and 1 labourer will fix 100 tiles in 1.25 hours:

Labour cost per hour (2 × £8.00) + (1 × £6.50) = £22.50

Cost per 100 tiles: 1.25 × £22.50	£28.13
	£107.40
Profit and oncost 20%	£21.48
Cost per 100 tiles	£128.88
Rate per m²: £128.88/2.37 m²	£54.34

Example 15.19

> Quarry tiles skirting; butt joints; including 100 mm bed of cement, sand mortar 1:3; jointing and pointing with cement, sand mortar 1:3; to floors; 1500 high; 150 × 150 × 16 mm units; round top, cove base. m

Materials

Skirting tiles cost £1.02 each

Length covered by 100 tiles: 100 × 0.154 = 15.40 m

Cost of 100 tiles: 100 × £1.02	£102.00
Unload and stack 100 tiles in 15 minutes: 0.25 × £6.50	£1.63
Mortar bed for 100 tiles at 10 mm thick: 0.0237 m³ at £65.32/m³	£1.55
Mortar grout: say 0.005 m³ at £65.32/m³	£0.33
	£105.50
Wastage 5%	£5.28

Labour

Squad of 2 tradesmen and 1 labourer will fix 100 tiles in 2.25 hours:

Labour cost per hour (2 × £8.00) + (1 × £6.50) = £22.50

Cost per 100 tiles: 2.25 × £22.50	£50.63
	£161.40
Profit and oncost 20%	£32.28
Cost per 100 tiles	£193.68
Rate per m²: £193.68/15.40 m	£12.58

16

Glazing

SMM7 Sections L40

Introduction

The use of glass in the building industry has increased greatly over the last few decades. The versatility of flat glass – the generic term used to describe unbowed glass used in the glazing of windows and doors – has enabled it to be used in many ways, which ultimately enhance the quality of life. Glass has been adapted to reduce both heat loss and noise with double and triple glazed units. Laminated and toughened glass offers greater safety than ordinary sheet and float glass; laminated glass gives the added advantage of increased security protection. Wired glass prevents the spread of flames, while solar control glasses (specially coated and tinted) help reduce the effects of radiant heat from the sun. Patterned and obscured glass afford a greater degree of privacy together with a decorative quality.

There are also numerous materials for the fixing of glass to frames, from traditional linseed oil and whiting putty, timber beads screwed or nailed to frames, to modern sealants, non-setting compounds, mastic tapes and so on. We shall, however, be concentrating on the basic methods by which glazing is fixed to frames.

Estimating for glazing

In order to estimate the costs of a particular item of work accurately, it is necessary to have details of the following:

- Glass type, quality, thickness and normal maximum sizes of sheets
- Type of surround to which glass is to be fixed, e.g. wood, aluminium, uPVC
- Glazing compounds, beads, etc. and their suitability for work in connection with a particular surround.

Wastage varies dependent on whether glass is bought pre-cut to the required size, or in large sheets, which are cut to size on site, or in the contractor's yard. Typical wastage figures are as follows:

Pre-cut glass	5%
Uncut glass	10%
Glazing compounds and sundries	10%

Under the rules laid down in SMM7 for glazing, standard plain glass is grouped into two main categories of pane area: less than 0.15 m^2, and from 0.15 to 4 m^2. Non-standard plain glass, that is any glass (other than special glass) which is drilled, brilliant cut or bent, or is 10 mm thick or more, or is in panes greater than 4 m^2, gives rise to another category of pane area: more than 4 m^2.

The variance in pane sizes within the category 0.15 to 4 m^2 can be great. Therefore the estimator will assess the average pane size and allow outputs accordingly. Although the

Table 16.1 *Approximate quantities of glazing compound for back and front fixing*

Pane sizes: adapted SMM7 categories (m^2)	Wood surrounds* (kg/m)	Metal surrounds* (kg/m^2)
≤0.15	3.00	4 00
>0.15 to 0.5	2.00	2.66
>0.5 to 1	1.00	1.33
>1	0.75	1.00

* For back putty only, 33.33 per cent of above weights should be used.

estimator may choose to determine his own categories and relate the relevant outputs, etc. to those categories, it is convenient in our case to adopt the categories used in SMM6.

Table 16.1 shows typical requirements for glazing compound. Table 16.2 shows outputs for a tradesman for cutting and glazing.

Types of glass

Clear sheet glass

Sheet glass, because of the nature of its manufacture, never has two surfaces perfectly parallel or flat, it thus distorts transmitted and reflected light. This glass is manufactured in thicknesses of 2, 3, 4, 5 and 6 mm (the 2 mm thickness is not recommended for general glazing). Clear sheet glass has generally been superseded by float glass.

The quality of the glass used depends on the class of work desired. BS 952 recommends three qualities:

Ordinary glazing quality OQ
Selected glazing quality SQ
Special selected quality SSQ

Table 16.2 *Tradesman's average outputs for cutting and glazing (full back and front fixing) (hours/m^2)*

Pane size (m^2)	Clear sheet ≤5 mm thick: Cut	Clear sheet ≤5 mm thick: Glazing Wood	Clear sheet ≤5 mm thick: Glazing Metal	Obscured, patterned or float ≤5 mm thick: Cut	Obscured, patterned or float ≤5 mm thick: Glazing Wood	Obscured, patterned or float ≤5 mm thick: Glazing Metal	Clear sheet, roughcast, polished plate >5 mm but ≤6 mm thick: Cut	Clear sheet, roughcast, polished plate >5 mm but ≤6 mm thick: Glazing Wood	Clear sheet, roughcast, polished plate >5 mm but ≤6 mm thick: Glazing Metal	Polished plate, float 10–12 mm thick and 6 mm wired glass: Cut	Polished plate, float 10–12 mm thick and 6 mm wired glass: Glazing Wood	Polished plate, float 10–12 mm thick and 6 mm wired glass: Glazing Metal
≤0.15	0.45	1.10	1.20	0.55	1.20	1.35	0.60	1.25	1.40	0.70	1.50	1.65
>0.15 to 0.5	0.35	0.80	0.90	0.45	0.90	1.00	0.50	0.95	1.05	0.55	1.20	1.35
>0.5 to 1	0.25	0.60	0.65	0.30	0.65	0.70	0.35	0.75	0.85	0.45	1.00	1.10
>1	0.20	0.45	0.50	0.25	0.50	0.55	0.30	0.60	0.65	0.35	0.80	0.90

Glazing with bradded wood beads	deduct 10%
Glazing with screw beads to wood	add 15%
Glazing with metal screw beads to metal	add 20%
Hacking out glass and preparing rebates to receive new glass	0.2 hours/m
Curved cutting, depending on type of glass and thickness	1.5–2.5 hours/m

Float glass

This is a truly flat glass, giving undistorted vision. It is formed by floating a continuous ribbon of molten glass onto liquid metal at a controlled rate and temperature. This glass is manufactured in general glazing quality and selected quality in thicknesses of 3, 4, 5, 6, 10 and 12 mm.

Polished plate glass

Polished plate glass has both surfaces ground and polished and in parallel, giving undistorted vision. It has generally been superseded by float glass, which has the same qualities at reduced cost. This glass is manufactured in thicknesses of 5, 6, 10 and 12 mm, although it can be manufactured up to 38 mm thick.

Patterned glass

Patterned glass is a rolled glass. During manufacture a pattern is formed on one face of the glass while the other face is left flat. The pattern makes the glass partially or totally obscure. This glass is manufactured in thicknesses of 3 and 5 mm.

Roughcast glass

Roughcast glass is a translucent glass which has a textured surface introduced by the rollers during manufacture. It is manufactured in thicknesses of 5, 6 and 10 mm.

Wired glass

Wired glass has wire mesh incorporated during the rolling process. Wired glass may have a cast finish or be subsequently polished. It is manufactured in thicknesses of 6 mm polished or 7 mm cast.

Laminated glass

This glass consists of a sandwich construction of two or more sheets of ordinary glass, between which are bonded interlayer(s) made of resilient plastic material. It is manufactured in thicknesses of 4.4, 5.4, 6.1, 6.4, 6.8 and 8.8 mm.

Toughened glass

Toughened glass is formed by heat-treating clear, patterned and other types of glass (wired glass cannot be treated in this manner). It is manufactured in thicknesses of 4, 5, 6 and 10 mm.

Example 16.1

4 mm OQ clear sheet glass to 20 mm wood rebates with putty; panes 0.15 to 4 m^2 (average pane sizes within category 0.15–0.5 m^2).	m^2

Materials	
Glass delivered to site per m^2	£18.40
Putty: 2 kg at £16.25 per 25 kg	£1.30
Glazing sprigs: say	£0.06
	£19.76
Allow waste (pre-cut glass) 5%	£0.99
	£20.75
Labour	
Tradesman 0.8 hours at £8.00	£6.40
	£27.15
Profit and oncost 20%	£5.43
Rate per m^2	£32.58

Example 16.2

4 mm OQ clear sheet glass to 20 mm metal rebates with glazing compound; not exceeding 0.15 m^2. 30 Nr. m^2

Materials	
Glass delivered to site per m^2	£18.40
Compound: 4 kg at £17.80 per 25 kg	£2.85
Steel sash clips: say	£0.16
	£21.41
Allow waste (uncut glass) 10%	£2.14
	£23.55
Labour	
Tradesman 1.65 hours at £8.00	£13.20
	£36.75
Profit and oncost 20%	£7.35
Rate per m^2	£44.10

Example 16.3

4 mm patterned glass to 20 mm wood rebates with putty; panes 0.15 to 4 m^2 (average pane sizes within category 0.15–0.5 m^2). m^2

Materials	
Glass delivered to site per m^2	£19.10
Compound: 2 kg at £16.25 per 25 kg	£1.30

Glazing sprigs: say	£0.06
	£20.46
Allow waste (uncut glass) 10%	£2.05
	£22.51
Labour	
Tradesman 1.35 hours at £8.00	£10.80
	£33.31
Profit and oncost 20%	£6.66
Rate per m^2	£39.97

Example 16.4

4 mm float glass to 20 mm wood rebates with bradded wood beads; panes 0.15 to 4 m^2 (average pane sizes within category 0.5–1 m^2). m^2

Materials	
Glass delivered to site per m^2	£18.40
Compound: 0.33 kg at £16.25 per 25 kg	£0.21
Brads (beads supplied by joiner, left loose): say	£0.07
	£18.68
Allow waste (uncut glass) 10%	£1.87
	£20.55
Labour	
Tradesman 1.35 hours at £8.00	£10.80
	£31.35
Profit and oncost 20%	£6.27
Rate per m^2	£37.62

Sealed multiple glazed units

Double or multi glazed units are factory made using two or more panes. The panes are not necessarily of the same type; a selection can be made from the flat glasses, such as float, patterned or wired. The panes are separated by special profiled spacers, ranging from 6 to 20 mm wide. The perimeter is sealed using an impervious, moisture resistant, polysulphide material. The enclosed space is filled with dry air or an inert gas. Double or multi glazed units are purpose made or in stock sizes for standard windows.

Example 16.5

Pilkington Insulight double glazing units; 2 nr 4 mm float glass with 6 mm air space; to wood rebates with general purpose butyl strip compression joint sealant and wood beads, screwed; 1830 × 1830 mm. Nr

Materials	
Unit size 1830 × 1830 mm delivered to site	£183.00
Butyl strip 2 × 4 × 1830 = 14.64 mm at £8.90 per 10 m	£13.03
	£196.03
Allow waste 5%	£9.80
	£205.83
Labour	
Tradesman 3 hours at £8.00	£24.00
	£229.83
Profit and oncost 20%	£45.97
Rate per unit	£275.80

Roof and wall glazing

Patent glazing systems

Patent glazing systems form a non-load-bearing glazed vertical or pitched external envelope to buildings. They consist generally of a framework of extruded aluminium alloy or PVC clad glazing bars, which support single glazing panes or double glazing units. Each system incorporates its own standard or purpose made flashing, trims, gutters, expansion joints, etc. No pricing examples are given of this form of construction, as individual project designs would be priced by a specialist company and included in a building contract as a nominated subcontract. Of the three types of roof and wall glazing described in this section, patent glazing is normally the cheapest.

Curtain walling systems

Curtain walling systems again provide a glazed external envelope to buildings in the form of a lightweight cladding attached to the structural framework of a building. The basic concept of curtain walling is a series of vertical mullions fixed to the structural frame by connectors which support the cladding's own deadweight and the imposed wind loads. Horizontal transoms form the connecting framework, to which are fixed glass panes or units and infill panels. As for patent glazing, no pricing examples are given of this form of construction, as individual project designs would be priced by a specialist company and included in a building contract as a nominated subcontract.

Structural glazing systems

Structural glazing is a form of curtain walling construction incorporating specially developed high performance silicone materials, by which glass panes or units are adhered to non-structural mullions and transoms. These are in turn fixed to the structural framework in a manner similar to curtain walling. The effect of a glass-only facade can be achieved using this type of system. Again no examples of price build-up are given owing to the specialist nature of the construction.

17

Decorative papers and painting

SMM7 Sections M52, M60

Painting and clear finishing

In the determination of the covering capacities of materials and the labour outputs for decoration work, the estimator is faced with more variables than with most other trades. For example, walls may vary greatly in porosity or texture. Therefore average coverages and outputs are given in Tables 17.1 and 17.2.

Under the coverage rules of SMM7, the preparation of surfaces and the rubbing down between coats with glass-, emery- or sand-paper are deemed to be included. Any other preparatory work or treatment work required, such as wire brushing, should be detailed in the specification or in bill items. A minimum preparation for sanding down is therefore taken in all items.

The labour outputs and covering capacities given in the tables are for general surfaces over 300 mm girth. Therefore the figures have to be adjusted, for example, where working to narrower girths or to window frames. Table 17.3 lists percentage adjustments for labour and materials. The covering capacities decrease, resulting in an increase in the costs of materials. Similarly, because of the increased difficulty of the work, the labour outputs decrease and thus the labour costs increase.

It is generally recognized that overheads incurred in the running of a painting and decorating company are not as great as those of a general builder. This can be attributed to a number of factors, namely fewer staff to administer, smaller premises to rent or buy, less investment in plant, and so on. In the examples that follow we allow 15 per cent for profit and oncost.

Table 17.1 *Covering capacities per 100 m^2 for decoration work*

Material	Surface	Quantity required to cover 100 m^2 (litres except as given)
Primer	Wood	8.0
	Metal	6.5
	Plaster	8.3
	Brick	12.5
	Concrete	9.0
Undercoat	Wood	7.2
	Metal	7.2
	Plaster	7.2
	Brick	7.2
	Concrete	7.2
Finishing coat	Wood	8.3
	Metal	8.3
	Plaster	8.3

	Brick	8.3
	Concrete	8.3
Emulsion	Plaster	8.3
Varnish size	Wood	7.0
Varnish	Wood	5.6
Knotting	Wood	0.75
Stopping (putty)	Wood	2.5 kg
Sandpaper	Preparing	8 sheets
Masonry sealer	Roughcast	25
Masonry paint	Roughcast	29

Table 17.2 *Labour outputs for decoration work*

Operation and surface	Tradesman's hours per 100 m^2
Washing, rubbing down existing gloss	7.0
Rubbing down wood	4.5
Wire brushing:	
Roughcast	7.0
Concrete	3.0
Metal	5.0
Knotting and stopping	6.0
Priming:	
Wood	5.8
Metal	5.0
Plaster	6.0
Brick	7.5
Concrete	5.8
Undercoat:	
Wood	5.0
Metal	4.7
Plaster	5.8
Brick	7.0
Concrete	5.7
Finishing coat:	
Wood	4.5
Metal	4.5
Plaster	4.5
Brick	6.7
Concrete	4.5
Size to wood	6.3
Varnish to sized wood	5.3
Emulsion:	
Unprimed plaster	6.3
Plaster	5.8
Paper	5.8
Concrete	6.0

Continued

Table 17.2 *Continued*

Operation and surface	Tradesman's hours per 100 m^2
Masonry paint sealer:	
Roughcast	7.3
Brick	7.0
Concrete	6.0
Masonry paint:	
Roughcast	7.0
Brick	6.8
Concrete	5.8

Table 17.3 *Labour and material cost increases on involved decoration work*

Description	Percentage increase in labour costs	Percentage increase in material costs
Isolated surfaces, girth ≤300 mm	30	10
Glazed windows and screens:		
Panes ≤0.1 m^2	50	20
Panes 0.1 to 0.5 m^2	25	10
Panes 0.5 to 1 m^2	10	nil
Panes >1 m^2	nil	less 10
Radiators, panel type:		
>300 mm girth	20	10
Isolated surfaces ≤300 mm girth	30	15
Railings, fences and gates, plain open type:		
>300 mm girth	10	5
Isolated surfaces ≤300 mm girth	30	10
Gutters, eaves:		
>300 mm girth	10	5
Isolated surfaces ≤300 mm girth	30	10
Services, pipes:		
>300 mm girth	10	5
Isolated surfaces ≤300 mm girth	40	15

Example 17.1

Preparing, knotting, stopping, one coat primer, two coats undercoat, one coat gloss finish to general surfaces of wood; over 300 mm girth. m^2

The cost is built up for 100 m^2.

Materials

Sandpaper preparing	8 sheets at £0.23	£1.84
Putty stopping	2.5 kg at £0.65	£1.63
Knotting	0.75 litres at £17.48	£13.11

Primer	8 litres at £3.95	£31.60
Undercoat (two coats)	14.4 litres at £2.90	£41.76
Finish	8.3 litres at £3.40	£28.22
		£118.16
Labour		
Preparing	4.5	
Stopping and knotting	6	
Primer	5.8	
Undercoat (two coats)	10	
Finish	4.5	
	30.8 hours at £8.00	£246.40
		£364.56

Brushes

A brush has an expected life of 100 hours. Say painter uses brush 66.6% of working hours. Then:

Painter hours a brush will last: $100 \times (100/66.66) = 150$

Brush costs say £6.70: cost per hour $£6.70/150 = 0.04$

Brush cost: 30.8 hours at £0.04	£1.23
	£365.79
Profit and oncost 15%	£54.87
Cost per 100 m^2	£420.66
Rate per m^2	£4.21

Example 17.2

Preparing, knotting, stopping, one coat primer, two coats undercoat, one coat gloss finish to general surfaces of wood; isolated surfaces <300 mm girth.	m

The cost is built up for 100 m^2.

Materials as Example 17.1	£118.16
Decreased covering capacity: add 10%	£11.82
Labour as Example 17.1	£246.40
Decreased covering capacity: add 10%	£24.64
Brushes as Example 17.1	£1.23
Decreased covering capacity: add 10%	£0.12
Cost per 100 m^2	£402.37
Rate per m^2	£4.02
Cost per metre for say 150 mm: £4.02 × (150/1000)	£0.60
Profit and oncost 15%	£0.09
Rate per m	£0.69

Example 17.3

Two coats emulsion paint to general surfaces of plaster walls; silk finish; over 300 mm girth.	m^2

The cost is built up for 100 m^2.

Materials		
Emulsion paint (two coats): 16.6 litres at £2.83		£46.98
Labour		
First coat	6.3	
Second coat	5.8	
	12.1 hours at £8.00	£96.80
		£143.78
Brushes		
Brush cost: 12.1 hours at £0.04		£0.48
		£144.26
Profit and oncost 15%		£21.64
Cost per 100 m^2		£165.90
Rate per m^2		£1.66

Example 17.4

Preparing, one coat primer, two coats undercoat, one coat gloss finish to services; pipes; isolated surfaces <300 mm girth.	m

The cost is built up for 100 m^2.

Materials as Example 17.1		£118.16	
Deduct knotting and stopping	£14.74		
Deduct 1.5 litres priming	£5.93	£20.67	£97.49
Decreased covering capacity: add 15%			£14.62
Labour as Example 17.1		£246.40	
Deduct knotting and stopping 6 hours	£48.00		
Deduct priming 0.8 hours	£6.40	£54.40	£192.00
Decreased covering capacity: add 40%			£76.80
Brushes as Example 17.1		£1.23	
Deduct knotting and stopping 6 hours	£0.24		
Deduct priming 0.8 hours	£0.03	£0.27	£0.96

Decreased covering capacity: add 40%	£0.38
Cost per 100 m^2	£382.25
Rate per m^2	£3.82
Cost per metre for say 150 mm: £3.82 × (150/1000)	£0.57
Profit and oncost 15%	£0.09
Rate per m	£0.66

Example 17.5

Two coats clear varnish to general surfaces of wood doors; over 300 mm girth.	m^2

The cost is built up for 100 m^2.

Materials		
Varnish size:	7 litres at £4.40	£30.80
Varnish (two coats):	11.2 litres at £5.74	£64.29
		£95.09
Labour		
Size coat	6.3	
Varnish coats	10.6	
	16.9 hours at £8.00	£135.20
		£230.29
Brushes		
Brush cost:	16.9 hours at £0.04	£0.68
		£230.97
Profit and oncost 15%		£34.65
Cost per 100 m^2		£265.62
Rate per m^2		£2.66

Example 17.6

Wire brushing surfaces of existing roughcast walls; one coat masonry paint sealer; two coats masonry paint; to general surfaces over 300 mm girth.	m^2

The cost is built up for 100 m^2.

Materials		
Wire brushes:	2 at £4.80	£9.60
Masonry sealer:	25 litres at £6.30	£157.50
Masonry paint (two coats):	58 litres at £6.55	£379.90
		£547.00

Labour		
Wire brushing	7	
Sealer coat	7.3	
Masonry paint coats	14	
	28.3 hours at £8.00	£226.40
		£773.40
Brushes		
Brush cost:	28.3 hours at £0.04	£1.13
		£774.53
Profit and oncost 15%		£116.18
Cost per 100 m²		£890.71
Rate per m²		£8.91

Decorative papers

Decorative papers are coverings applied to internal walls or ceilings with glue or paste. There are plain and decorative papers, and fabrics and plastics on paper backings. Papers are sold in rolls or pieces to a standard size of 10 m length by 530 mm width. There are instances where paper sizes may differ from this standard, which would have to be accounted for in the development of rates.

A decision on the exact pattern of paper to be used may not have been made at the time of producing bills of quantities. However, if the quality or price range is known then a prime cost sum can be calculated. The prime cost sum will include the supply and delivery of the required paper to the site.

Table 17.4 shows wastage values for decorative papers. Table 17.5 gives labour outputs for straight cutting, pasting and hanging papers.

Table 17.4 *Average wastages for decorative paper*

Description	Percentage waste
Lining paper to walls	7.5
ceilings	5.0
Plain paper to walls	10.0
ceilings	7.5
Pattern paper to walls	15.0
ceilings	12.5
Borders and strips	7.5

Table 17.5 *Labour outputs for straight cutting, pasting and hanging decorative papers*

Description	Tradesman's hours per roll
Strip old paper from walls	0.53
ceilings	0.64
Clean down, stop walls	0.27
ceilings	0.32
Glue size wall or ceilings	0.37
Light papers to walls	1.00
ceilings	0.80
Heavy papers to walls	1.33
ceilings	1.00
Borders 40 mm wide	20 m/hour

Example 17.7

Stripping existing decorative paper, stopping, glue size and hang heavy patterned decorative papers (PC sum £6.80/roll) to plaster surfaces of walls.	m^2

The cost is built up for one roll.

Materials		
Sand paper preparing	0.5 at £0.23	£0.12
Stopping	0.1 kg at £0.97	£0.10
Glue size	1 litre at £0.68	£0.68
Decorative paper	1 roll at £6.80	£6.80
		£7.70
Labour		
Strip old paper	0.53	
Prepare and stop	0.27	
Glue size	0.37	
Hang paper	1.33	
	2.50 hours at £8.00	£20.00
		£27.70
Profit and oncost 15%		£4.16
Cost to hang 1 roll (5.3 m^2)		£31.86
Rate per m^2		£6.01

18

Drainage

SMM7 Sections R12, R13

Excavating drainage trenches

The SMM7 unit of measurement for excavating drainage trenches is the metre, unlike groundworks (see Chapter 5) where the unit of measure is the cubic metre. The costs for excavation of drainage trenches, disposal of surplus excavated material, and earthwork support can be developed using the same labour constants as those in Chapter 5. However, the m^3 rates for excavation have to be adjusted by applying the volume of a metre run of trench to establish a metre rate.

The estimator has to cost certain items deemed to be included by the coverage rules of SMM7. These items are for earthwork support, consolidation of trench bottoms, trimming excavations, filling with and compaction of general filling materials, and disposal of surplus excavated material.

As no width of trench is required to be given in the item description, we have to establish widths suitable for the purpose of laying pipes. The following are acceptable trench widths for pipes up to 200 mm diameter:

Depth of trench (m)	≤1	≤2	≤4	≤6
Width of trench (mm)	550	600	750	900

Trench widths for pipes over 200 mm diameter can be established by adding the diameter of the pipe to the appropriate trench width above.

Hand excavation of pipe trenches

There is no longer a requirement in SMM7 to state depth ranges in items for excavation of pipe trenches; it is necessary only to give the average depth of excavation in stages of 250 mm. However, as covered in Chapter 5, the cost of excavating by hand is calculated in 1.5 m stages, that being the recognized maximum height that a man can throw a spade full of excavated material.

We shall first determine the cost per m^3 to excavate pipe trenches, fill with and compact filling materials, and dispose of surplus excavated material, for each 1.5 m stage or throw, i.e. up to 1.5 m deep, 1.5 to 3 m deep, and 3 to 4.5 m deep. The costs are then used in subsequent examples.

Depth stage up to 1.5 m

	hrs/m^3	
Excavate and get out	2.5	
Fill, compact, dispose of surplus	1.5	
Labour output	4	
Cost per m^3 at £6.50/hour		£26.00

Depth stage 1.5 to 3 m

	hrs/m^3	
Excavate and get out	2.5	
Extra, to throw from 3 m to 1.5 m	1	
Fill, compact, dispose of surplus	1.5	
Labour output	5	
Cost per m^3 5.0 at £6.50		£32.50

Depth stage 3 to 4.5 m

	hrs/m^3	
Excavate and get out	2.5	
Extra to throw 3 m to 1.5 m	1	
Extra to throw 4.5 m to 3 m	1	
Fill, compact, dispose of surplus	1.5	
Labour output	6	
Cost per m^3 6.0 at £6.50		£39.00

Example 18.1

> Excavating trenches for pipes not exceeding 200 mm nominal size; trimming excavations; earthwork support; consolidation of trench bottoms; filling with and compaction of material arising from excavations; disposal of surplus excavated material by removing from site; average 1.25 m deep. m

Volume of trench: 1.00 × 0.60 × 1.25 per metre of trench				0.75 m^3
Cost per m^3 for depth stage up to 1.5 m:				
Cost per metre:	£26.00	0.75	m^3	£19.50
	l	*h*	*m^2*	
Trimming excavations: m^2 per metre of trench	2.00	0.60	1.20	
Cost per metre at 0.08 labourer hours/m^2:				£0.62
Consolidation of trench bottoms: m^2 per metre of trench				
	l	*w*	*m^2*	
	1.00	0.60	0.60	
Cost per metre at 0.08 labourer hours/m^2				£0.31
	l	*h*	*m^2*	
Earthwork support to 2 sides of excavations:	2.00	1.25	2.50	
Cost of earthwork support (see Chapter 5)				£15.88
				£36.31

Profit and oncost 20%	£7.26
Rate per metre	£43.57

Example 18.2

Excavating trenches for pipes not exceeding 200 mm nominal size; trimming excavations; earthwork support; consolidation of trench bottoms; filling with and compaction of material arising from excavations; disposal of surplus excavated material by removing from site; average 2.5 m deep.	m

Volume of trench: 1.00 × 0.75 × 2.50 per metre of trench				1.88 m^3
Volume of trench in stage to 1.5 m: 1.13 m^3 per metre of trench				1.13 m^3
Volume of trench in stage 1.5 to 3 m: 0.75 m^3 per metre of trench				0.75 m^3
Cost per m^3 for depth stage up to 1.5 m:				
Cost per metre:	£26.00	1.13 m^3		£29.25
Cost per m^3 for depth stage 1.5 to 3 m:				
Cost per metre: £26.50 × 0.75	£32.50	0.75 m^3		£24.38
	l	*h*	*m²*	
Trimming excavations: m^2 per metre of trench	2.00	2.50	5.00	
Cost per metre at 0.08 labourer hours/m^2:				£2.60
Consolidation of trench bottoms: m^2 per metre of trench				
	l	*w*	*m²*	
	1.00	0.75	0.75	
Cost per metre at 0.08 labourer hours/m^2				£0.39
				£88.37
Profit and oncost 20%				£17.67
Rate per metre				£106.04

Machine excavation of pipe trenches

The costs for excavation of pipe trenches by machine need not be determined by the same depth ranges as those used in hand excavation. Machine excavation can be calculated using deeper depth ranges, so limiting the number of range costs to be calculated. The only restriction for a machine with regard to depth is the maximum reach of a particular machine; therefore the correct machine has to be chosen to cope with a required depth. Machine outputs will decrease as the excavation deepens; Example 18.3 allows a reasonable average output per hour for each range.

Example 18.3

Excavating trenches for pipes not exceeding 200 mm nominal size; trimming excavations; earthwork support; consolidation of trench bottoms; filling with and compaction of material arising from excavations; disposal of surplus excavated material by removing from site; average 2.5 m deep.	m

An excavator/loader will be used with a backactor of capacity 0.22 m^3.		
Hire of machine for 39 hour week		£592.00
Driver 39 hours at £6.85 per hour		£267.15
Labourer 39 hours at £6.50		£253.50
Excavator cost per week		£1112.65
Hours per week	39	
Less standing time 10% – say	4	
Operating hours per week	35	
Hourly cost		£31.79
Cleaning and maintaining at 0.5 hours per 8 hour day/39 at £6.85/hr		£0.43
Fuel per hour: 2.5 litres at £0.56		£1.40
Other consumables, oil, etc.: say of fuel cost		£0.14
Insurance: say		£0.18
Hourly cost of excavator		£33.94

Outputs

Using a backactor of capacity 0.22 m^2:
8.0 m^3/hour should be excavated to a depth of 2 m
7.2 m^3/hour should be excavated from 2 to 4 m

Cost to excavate 1 m^3 up to 2 m	£4.24 per hour
Cost to excavate 1 m^3 from 2 to 4 m	£4.71 per hour

Volumes	*l*	*w*	*d*
Trench total: 1.00 × 0.75 × 2.50	1.00	0.75	2.50
Volume is			1.88 m^3
Trench to 2.00 m deep: 1.00 × 0.75 × 2.00	1.00	0.75	2.00
Volume is			1.50 m^3
Remainder of trench:	1.00	0.75	0.50
Volume is			0.38 m^3

Costing

Cost to excavate trench to 2 m:				£6.36
Cost to excavate trench from 2 to 4 m				£1.77
	l	*h*	*m^2*	
Trimming excavations: m^2 per metre of trench	2.00	2.50	5.00	

Cost per metre at 0.08 labourer hours/m^2:				£2.60
Consolidation of trench bottoms: m^2 per metre of trench				
	l	*h*	*m^2*	
	1.00	0.75	0.75	
Cost per metre at 0.08 labourer hours/m^2				£0.39
	l	*h*	*m^2*	
Earthwork support to 2 sides of excavations:	2.00	2.50	5.00	
Cost of earthwork support (see Chapter 5)				£31.75
Fill and compact excavated material, dispose of surplus: 1.5 hours/m^3				
Cost per metre: vol. × hrs × lab rate				£18.28
				£61.15
Profit and oncost 20%				£12.23
Rate per metre				£73.38

The amount of fill must be adjusted if a bed and a surround of a particular material are specified. The quantity of fill will decrease and the quantity of material to be disposed of will increase.

* Driver Class N – £23.01/week 50 p/hour – Assumes front loader capacity is over 0.58 m^3 ($\frac{3}{4}$ cubic yard).

Pipework: vitrified clay and uPVC

The examples that follow deal with pipework that is most commonly used in drainage systems where low volumes of waste material and water have to be handled, e.g. house drainage up to its connection with the main sewer. The two main types of pipe used in this situation are vitrified clay and uPVC. These pipes are very resilient and are relatively easy to install. Cast iron pipes are very rarely used nowadays in underground drainage mainly due to the high material and installation costs.

Example 18.4

100 mm diameter vitrified clay pipes in trenches; push fit polypropylene flexible coupling joints.	m

Materials

Pipes in standard lengths of 1.6 m delivered: each	£3.48
Take 1 polypropylene coupling per length	£1.25
Joint lubricant: 1 kg per 50 m pipework at £2.85/kg	
Cost per joint = (cost per kg/50) × 1.6	£0.09
	£4.82
Deduct full load discount 17.50% (including fittings and sundries)	£0.84
	£3.98
Allow wastage 5%	£0.20
	£4.18

Labour

Unload and stack 1 labourer 20 lengths/hour at £6.50: per length	£0.33
Pipelayer and labourer lay one length every 5 minutes: (£6.80 + £6.50)/(5/60)	£1.11
	£5.61
Profit and oncost 20%	£1.12
Cost per 1.6 m length	£6.73
Rate per metre	£4.21

Example 18.5

Extra over 100 mm diameter vitrified clay pipes for single branches.	Nr

Materials

Cost of branch	£5.10
Take 2 polypropylene couplings per branch	£2.50
Cost of joint lubricant previous example × 2	£0.18
	£7.78
Deduct full load discount 17.50% (including fittings and sundries)	£1.36
	£6.42
Allow wastage 2.5%	£0.16
Labour	
Unload and stack, labourer 40 branches/hour at £6.50: per branch	£0.16
Pipelayer and labourer lay one branch every 15 minutes: (£6.80 + £6.50)/(15/60)	£3.33
	£10.07
Deduct length of pipe occupied by branch – 0.75 m × Ex 18.4	£3.16
Extra over cost for branch	£6.91
Profit and oncost 20%	£1.38
Extra over rate per branch	£8.30

Example 18.6

Extra over 100 mm diameter vitrified clay pipes for bends.	Nr

Materials

Cost of bend	£2.28
Take 1 polypropylene coupling per bend	£1.25
Cost of joint lubricant as Example 18.4	£0.09
	£3.62
Deduct full load discount 17.50%	£0.63
	£4.25
Allow wastage 2.5%	£0.11
	£4.36

Labour	
Unload and stack, labourer 40 bends/hour at £6.50: per bend	£0.16
Pipelayer and labourer lay one bend every 10 minutes: (£6.80 + £6.50)/(10/60)	£4.47
Deduct length of pipe occupied by branch – 0.30 m × Ex 18.4	£0.83
Extra over cost	£3.64
Profit and oncost 20%	£0.73
Extra over rate per bend	£4.37

Example 18.7

150 mm diameter vitrified clay pipes in trenches; push fit flexible socket and spigot joints.	m

Materials	
Pipes in standard lengths 1.5 m delivered site	£11.30
Cost of joint lubricant as Example 18.4	£0.09
	£11.39
Deduct full load discount 17.50%	£1.99
	£9.40
Allow wastage 5%	£0.47
Labour	
Unload and stack, labourer 20 lengths/hour at £6.50	£0.33
Pipelayer and labourer lay one length every 8 minutes: (£6.80 + £6.50)/(8/60)	£1.77
	£11.97
Profit and oncost 20%	£2.39
Cost per 1.5 m length	£14.36
Rate per metre	£9.57

Example 18.8

110 mm diameter uPVC pipes in trenches; ring seal joints.	m

Materials	
Pipes in standard lengths of 3 m delivered site, single socket	£22.00
Joint lubricant: 1 kg per 50 m pipework at £2.85/kg	
Cost per joint = (cost per kg/50) × 3.00)	£0.17
	£22.17
Deduct discount of 15.00%	£3.33
	£18.85
Allow wastage 3%	£0.57
	£19.41

Labour	
Unload and stack, labourer 15 lengths/hour at £6.50: per length	£0.43
Pipelayer and labourer lay one length every 7 minutes: (£6.80 + £6.50)/(7/60)	£1.55
	£21.40
Profit and oncost 20%	£4.28
Cost per 3 m length	£25.67
Rate per metre: £13.49/3	£8.56

Beds and surrounds

Example 18.9

Plain *in situ* concrete: grade C20P, 20 mm aggregate; beds and coverings; 400 × 400 mm to 100 mm diameter pipes.	m

Materials	*l*	*w*	*t*	*m³*	
Volume of concrete and pipe per metre of pipe	1.00	0.55	0.40	0.220	
	π	*r*	*l*	*m³*	
Volume of pipe with wall thickness of 10 mm	3.14	0.06	1.00	0.011	
Volume of concrete only				0.209	
Cost of 1 m^3 concrete grade C20P, 20 mm aggregate, including mixing, transporting and placing per m^3					£65.00
Profit and oncost 20%					£13.00
Cost per m^3					£78.00
Rate per metre					£16.28

It is taken that concrete will be poured to the sides of the trenches; therefore no formwork has been included in this example. It should, therefore, be noted that as the minimum width of trench is 550 mm, concrete used will be $0.55 \times 0.40 \times 1.00 = 0.22\,m^3$. The example which follows shows the cost effect of using formwork in the above situation.

Example 18.10

Plain *in situ* concrete: grade C20P, 20 mm aggregate; beds and coverings; 400 × 400 to 100 mm diameter pipes.	m

Materials	*l*	*w*	*t*	*m³*	
Volume of concrete and pipe per metre of pipe	1.00	0.40	0.40	0.160	
	π	*r*	*l*	*m³*	
Volume of pipe with wall thickness of 10 mm	3.14	0.06	1.00	0.011	
Volume of concrete only				0.149	
Cost of 1 m³ concrete grade C20P, 20 mm aggregate, including mixing, transporting and placing per m³					£65.00
Cost of concrete per metre					£9.67
		l	*h*	*m²*	
Formwork* to 1 m length of beds and coverings at £28.60 per m²	2	1.00	0.40	0.80	
					£22.88
					£32.55
Profit and oncost 20%					£6.51
Rate per metre					£39.05

* Includes erection and sliding on completion.

19
Electrical work

SMM7 Sections V, P31

Introduction

An electrical system is designed by an electrical engineer to comprise a number of manufactured pieces of equipment and fittings such as switches, socket outlets, cable outlets, light fittings, etc., all connected by given sizes of electrical cable. There is a wide range of cable types and protective systems available each of which has to be identified separately in a bill of quantities. The work on site entails taking delivery, storing, setting or supporting in place and connecting the equipment, outlets and fittings. The electrical work on site concerns installation and not manufacture (unless one considers the adaptation of standard lengths of cable and their protective systems as manufacturing) therefore the processes involved in estimating tend to be relatively straightforward.

Primary switchgear

The primary switchgear and distribution equipment are the first items to be connected to the electricity authority's supply. In a domestic situation this will be the consumer unit. A large manufacturing plant might have a substation comprising transformers, fuses, metering, circuit breakers, etc. Example 19.1 illustrates the estimation for the installation of a cubicle switchboard to be installed in a light industrial unit.

Example 19.1

Cubicle switchboard fixed to concrete floor with M10 rawlbolts, bolt projecting type, complete with all labels, fuses, MCBs, time switches, contactors, etc. as shown on drawing nr 8706(E)01.	Nr

The unit weighs 450 kg, and therefore the total cost is that involved in taking delivery and placing in position. It is assumed that the unit will be installed in a large cupboard on the ground floor, at the end of a 30 m corridor from which access may be gained to the outside; approximate distance to lorry 10 m.

The constants used in this example are based on the hours per tonne expended by each of the operatives required to move the weight. It is assumed that two operatives are capable of carrying out the task involved. If a 1 tonne piece of equipment were being moved, not only would the hours based upon the constants be increased, but also three operatives might be needed to move the weight. The cost therefore rises exponentially in proportion to the weight. To move such equipment on an existing hard floor surface, 'skates' can be used or a low wheeled trolley can be manufactured for the purpose. 'Skates' are small metal plates approximately 200 × 100 × 30 mm high, each fitted with pairs of rollers. Four would be placed under the corners of the cabinet by simply levering up or jacking each corner in turn and then the whole thing rolled down the corridor into position. Turning corners has to be achieved by lifting and realigning the skates as they are not steerable. This can be time consuming.

Cost of unit delivered to site			£4000.00
M10 Rawlbolts	4	£1.60	£6.40
Cost of installation is calculated as follows:			
Mass of unit in kg	450		
	Hours/operation		
Unload from vehicle hr/tonne	1.30	0.59	
Jack up and place skates hr/tonne	0.80	0.36	
Place ramp at entrance		0.50	
Move externally, say, 10 m per 10 m	0.50	0.20	
Move internally per 10 m	0.60	0.80	
Turn twice per turn	1.60	1.40	
Line up to final position	1.00	0.50	
Jack down and remove skates	0.80	0.40	
Final positioning and bolt down	3.00	1.40	
		6.15	
		2.00	
Total time for 2 operatives		12.29	
Labourer's rate	£6.50		
Electrician's rate	£12.80		
Cost of installing		£19.30	£237.20
			£4243.60
Profit and oncost 20%			£848.72
Rate each			£5092.32

Although rawlbolting might be considered 'Builder's work in connection' with this item, fastening with proprietory fasteners is most commonly undertaken by the trades people involved.

Cabling

Cabling has three material components:

- The cable support, i.e. the cable tray or conduit
- The fixing of the cable tray or conduit to the structure
- The cable itself.

Within the light industrial unit of Example 19.1, the cable is carried from the switchgear fuseboard on a cable tray in the corridor and enters plastic oval conduit for the final run to termination, i.e. the switch, lighting outlet point or power outlet point.

Example 19.2

Standard duty, return edge galvanized cable tray 300 mm wide with bolted joints fixed to softwood joists.	m

	Hours	Hourly rate	
Galvanized tray per metre			£11.50
Labour fixing 0.6 hours/m at £7.65	0.6	£12.80	£7.68
			£19.18
Profit and oncost 20%			£3.84
Rate per metre			£23.02

Example 19.3

25 mm straight PVC oval conduit fixed with clips to blockwork.	
Lighting outlet points in ceiling void.	Nr
One way switch points.	Nr

It should be noted that although SMM7 requires the lengths of both conduit and cable to be stated, the above description is not unusual. It is necessary in these cases to determine from the drawings the likely lengths of both conduit and cable. It is assumed, in this example, that the following lengths for 10 lighting outlets and 10 switches have been determined from the drawings:

Total length of conduit	150 m
Total length of cable	900 m (one line, one neutral, one switch for two way switching and one earth plus tails at each fitting)

Thus the item measured is:

	total	Unit cost	Per 100 m
25 mm PVC oval conduit/100 m	150	£22.00	£22.00
1.00 mm single core PVC insulated cable/100 m	900	£6.00	£36.00
Clips for conduit: 2/metre at cost per 100	2	£4.68	£9.36
Pins for clips: 4/m at cost per 100	4	£2.00	£8.00
Cost for 100 metres			£75.36
Add waste 15.00%			£11.30
			£86.66
Cost per metre			£0.87

	Hours	All-in hourly	
Labour fixing conduit and drawing in cable/metre	0.6	£12.80	£7.68
			£8.55

Profit and oncost 20%			£1.71
Rate per metre			£10.26

In order to price the item in the bill of quantities it is necessary to take the product of the measured length and divide by the number of fittings:

Cost for 150 m of conduit	150	£10.26	£1538.40
Number of fittings	20		
Rate per fitting			£76.92

The installation of electrical fittings is a two-stage process involving a first and second fix. During the first fix the conduits, cables and back boxes for fittings are installed. This may involve preparatory work by the builder, which is described later. Following the finishing trades (plasterer, tiler, etc.), the electrician will return for the second fix to connect the wiring to the fitting and fix the fitting to the back box.

Example 19.4

13 amp, 1 gang, switched socket outlet to BS 1363:1967 with white plastic cover-plate and galvanized steel box to BS 4662:1970 for flush mounting plugged and screwed to brickwork.	Nr

Materials			
13 amp, 1 gang, switched socket outlet and box			£1.65
Pressed steel 35 mm back box			£0.55
Plugs and screws: say			£0.05
			£2.25
Add 5% waste			£0.11
Labour			
1st fix: fixing box	0.2	£12.80	£2.56
2nd fix: wiring socket and fixing coverplate to box	0.25	£12.80	£3.20
			£8.12
Profit and oncost 20%			£1.62
Rate per fitting			£9.75

Builder's work

This work is covered by Section P31 of SMM7. It comprises preparatory work, e.g. cutting chases in brick walls and forming holes, and is carried out by the builder for the electrician. The work will be undertaken by a tradesman or a labourer depending upon its complexity.

Example 19.5

Cutting or forming chase in brickwork for concealed single 25 mm straight PVC oval conduit.	m

Labourer cuts 25 × 25 mm chase in brickwork with mechanical hammer 0.125 hours/m at £5.30	0.13	£12.80	£1.60
Profit and oncost 20%			£0.32
Rate per metre			£1.92

Example 19.6

Cutting or forming sinking in brickwork for concealed 80 × 80 × 35 mm deep galvanized steel box for socket outlet point.	Nr

Labourer cuts 80 × 80 × 35 mm sinking	0.18	£12.80	£2.30
Profit and oncost 20%			£0.46
Rate per sinking			£2.76

20
Analogous rates

Introduction

Whether you are a builder or a surveyor, you will some day be confronted with the need to produce a rate for work carried out on site which has not been priced in a bill of quantities or a schedule of rates. Such a rate is termed an 'analogous rate' or a 'pro rata rate'. While not decrying the term 'pro rata', we note that it means 'proportionally'. The purpose of this chapter is to show that not all elements of a bill rate are proportioned directly, so the preferred term is 'analogous rate'. The OED defines analogous as 'similar, parallel (to)' and in a biological sense as 'similar in function but not necessarily in structure or position'. This latter definition is very close to the idea of an analogous rate.

This chapter deals with the general principles involved in the analysis of existing rates and the build-up or synthesis of analogous rates for several trades or work sections. The builder should have ready access to his estimator's original calculations and should have no difficulty working out what the new rate should be – from his point of view. This will not generally coincide with the view held by the client's quantity surveyor who, having no access to these original figures, will have to use the techniques described in this chapter to assess the validity of the builder's claim. This chapter has been written with the novice quantity surveyor in mind.

It is usual to require an analogous rate for work caused by something like a site instruction which has varied the original construction. This new rate therefore has to be worked out on the same basis as the rates in the bill or schedule. This means that certain elements in the rate will be the same as the original rates and others will differ.

The reason we wish to use a similar basis to the original rates for the analogous rates is simply that when the estimator priced the original bill he allocated the costs of many of the fixed items in his own particular way, and we must follow the same pattern or our new rates will be distorted. The quantity surveyor assessing new rates must be aware of the types of thing this involves. For example, the bill items for say external renderings may include only a portion of the total cost of scaffolding, or no allowance at all for scaffolding. In the first case the estimator may have included all of bringing scaffolding to site, erection, and taking down and removal in the bricklaying rates. Thus scaffolding for the roughcaster is only on the basis of 'cost while it is left a week or two longer'. In the second instance, the total cost of scaffolding is placed in the preliminaries and is not apportioned over the bill rates for measured work.

Elements of analogous rates

Generally the elements making up a rate for measured work are as follows:

- Materials
- Labour
- Plant
- Oncost
- Profit.

For each of these elements there are a number of factors which determine the level of cost which the estimator originally included in the bill rates; these must be analysed.

Material

The factors here are

- Quantity discounts
- Bulk buying or ordering
- Advance purchase.

The first two may seem much the same, but there are subtle differences! Put yourself in the builder's place. Order 1 tonne of OP cement and you will be quoted a price. Order 100 tonnes and the price per tonne will be a lot less. The same holds true for bulk buying or ordering, but in this case the builder places an order for 100 tonnes of OP cement in advance of getting the firm order for the work. He then 'draws' the OP cement in smaller quantities as he requires it. Not necessarily such a good price will be quoted.

Advance purchase is generally made when the builder has foreknowledge of price increases in basic materials which he knows will be utilized fairly soon. 'Fairly soon' is important because he is tying up capital or extending borrowing in the hope of making a larger profit in the reselling of the material. The additional profit must be greater than the cost of borrowing or the loss of interest on capital otherwise invested.

Labour

Particular considerations here are

- Outputs
- Bonus
- Weather
- Site management and organization.

These have all been discussed in Chapter 2, so there is no need to repeat the arguments here.

Plant

Here the estimator must consider

- Availability
- Suitability
- Level of maintenance
- Level of skill in use.

Oncost

Oncosts include head office overheads and site overheads.

Profit

The level of return required must be established.

Procedure

The procedure generally adopted to make up a rate analogous to a bill rale is as follows:

1. Establish the percentage for oncost and profit and reduce the original bill rate to net cost.
2. Isolate unknown factors by calculating the known part and deducting this from the net cost. Generally this means calculating the cost of materials, leaving labour and plant costs as the unknowns.
3. Of labour and plant costs, the easier to calculate will be the plant cost.
4. Then adjust the labour cost.
5. Work out the new plant cost.
6. Add materials costs.
7. Add oncost and profit.

Before launching into an expansion of these ideas it must be stressed that the new work for which the analogous rate is required must be comparable with the bill item which will form the basis for the calculation. It is no use trying to build up a new rate for machine excavation for a pipe trench using a rate for hand excavation of manholes. This is an extreme example, but serves as a warning that all factors have to be considered. Where, when and how is the new work to be carried out? If the answers match up with the bill item used, then go ahead.

Depending on the variation incorporated in the new item, one or all of the elements will change. For example, suppose the bill has an item for a half brick wall built in Blogg's heather facing brick, and a site instruction is issued to use Blogg's wire cut rustic buff facing brick at an increased cost of £50 per thousand. Obviously if the bricks are the same size, use the same mortar and have the same weight, then the only difference between the original and the analogous rate is the difference in cost for the bricks and the adjustment of oncost and profit on that difference. This is a fairly common problem, and will be treated more seriously as an example later in the chapter.

Assessment of oncost and profit

If we assume that the original estimator was reasonably consistent, then we can further assume that the percentages added for oncost and profit were the same for all major items.

In that case we must examine the bill and extract the rates for the principal items which carry the major part of the cost for the particular trade or work section with which we are dealing. We now carry out the process of building up our own net costs for these items. The analysis of the rates described in this step has to imitate as closely as possible the original estimator's mental process. This analysis must obviously include the rate(s) that we wish to change or to which our new rate(s) will be analogous. We subtract our net cost from the bill rate. The difference is expressed as a percentage of our net rate, and the percentage is noted.

Once we have a number of rates analysed in this way we should see a pattern emerging in the percentages. We can adjust our assumptions on labour costs, materials costs and plant costs in the light of the differences in these percentages, until we are satisfied that we have duplicated as nearly as possible what the original estimator did when he priced the bill or schedule. We should now have isolated oncost and profit, although labour, plant and material will still be only notional, that is, our guestimate. However, we should be in a position to reduce our bill rate to a net cost.

Isolation of unknown factors

At this stage we generally wish to find the cost of the materials used in the bill rate, simply because it is the information most readily available. Manufacturers and suppliers can be consulted at this stage as well as price books and price listings in the technical press. One must be aware that prices quoted may or may not include the cost of transport, loading and unloading, site handling and the possibility of bulk discounts. One must also be aware that the materials cost in the bill rate is the cost to the contractor when the work was carried out. This may seem obvious but needs emphasizing. The analogous rate may be calculated anything up to a year or more after the contract has been completed. Price lists or quotes at that date are no help in analysing a rate for work carried out so much earlier.

Deducting the materials cost from the net rate leaves us with the two unknowns of plant and labour. Of these, labour is the more problematical. We have the difficulty of knowing the level of wages and bonus allowed for; it is very difficult to duplicate an estimator's figures in respect of hourly labour costs and outputs. One can only guess at things like squad sizes, allowances for guaranteed week, overtime factors and travelling time. So the one to tackle is the plant figure.

Calculation of plant cost

The quantity surveyor will have some idea of the make-up of the contractor's organization; for example, the contractor may own some basic machinery but will hire large or specialized items. Plant costs as well as outputs can be obtained readily from plant hirers, and one can then assess the proportion of the plant cost in the bill rate fairly easily. Subtraction of this last sum gives the labour cost.

Adjustment of labour cost

This step marks the beginning of the synthesis of the new rate(s).

In some instances the labour cost can be directly proportioned to the cost in the bill rate. If the labour will vary significantly, it may be necessary to calculate a labour cost from new output figures, squad sizes and all-in hourly rates. Look at the examples for different ways to carry this out. To the adjusted labour cost is added the new plant cost.

Addition of new plant cost

Previous work will leave the quantity surveyor with lots of information on plant costs which can now be used to give a new plant cost. It may not always be necessary to change the plant cost at all, as illustrated in some examples later in the chapter.

Addition of new materials cost

Prices for new materials are added at this point. The sources are the same as for the analysis of the bill rates.

Addition of oncost and profit

Adding oncost and profit completes the analogous rate. However, the budding quantity surveyor should not feel that this is the end of the matter. Some hard bargaining might well follow when this new rate is proudly displayed to the contractor and his surveyor. Remember, they have all the original information at their fingertips!

Interpretation of bills and schedules

We are all aware that the construction industry is never as simple as we have just described – and never forgiving of foolish generalizations. The estimator may not have put all the oncost in the item rates; the profit may all be in the percentage added on at the end of the tender; the plant and scaffold may all be in the preliminary bill; and so on. The earlier chapters on preliminaries, tendering and so on should help you look at priced bills with the right ideas in mind. You should examine bills with the object of spotting front loading, net item rates, etc. and then you must be able to adjust the steps above accordingly.

Of course, there are other clues available to you. The fact that as a quantity surveyor you visit the site to carry out regular interim valuations means that you actually see what is going on; for example, you should note what general plant is kept on site. You can obtain much information from a good clerk of works. Look at the clerk of works log book and see which and what size of squads were or are on site. The clerk of works can tell you if any special provisions had to be made for some parts of the work.

Many contracts now require that the contractor prepares a *modus operandi* or plan of work: where he will use certain types of plant, the order in which work will be done, squad sizes, etc. The contractor may also be expected to provide critical path networks and keep these up to date on a weekly basis. These can identify, to the knowledgeable, where the contractor has used plant, varied squad sizes, etc. Earlier interim valuations with their list of materials on site should alert you to the size of loads being brought in and, for example, whether materials are being bought in bulk at a discount or favourable price.

All the clues are there; they require first to be seen, and second to be correctly interpreted.

Before continuing with a number of examples of analogous rates calculations, we will assume that analysis of principal rates has shown that the level of profit and oncost is 20 per cent.

Example 20.1

Bill rates: machine excavation

Excavate foundation trench not exceeding 2 m deep.	m^3	£4.04
Excavate foundation trench not exceeding 4 m deep.	m^3	£4.51
Basement excavation not exceeding 2 m deep.	m^3	£1.44

Variation account: machine excavation

Basement excavation not exceeding 4 m deep.	m^3	£1.61

It is assumed that the starting level of all excavations is within 0.25 m of existing ground level. The rate is arrived at by direct proportion. So here we have a true pro-rata rate. The proportional increase in the basement excavation from one depth stage to another is the same as for the trench excavation:
£1.44 × (4.51/4.04) = £1.607.

Example 20.2

Bill rates: for disposal of excavated material

(a) Deposit surplus spoil in heaps, wheeling not exceeding 100 m.	m^3	£1.10
(b) Deposit surplus spoil in heaps, wheeling not exceeding 150 m.	m^3	£1.36
(c) Spread and level surplus spoil, wheeling not exceeding 100 m.	m^3	£1.68

Variation account

Spread and level surplus spoil, wheeling not exceeding 200 m.	m^3

To wheel 50 metres: item (b) less item (a)	£0.26
To wheel a further 50 metres	£0.26
	£0.52
So spread and level and wheeling 200 metres = item (c) plus extra wheel	£1.68
	£2.20

In this example only part of the rate is proportioned – the wheeling. The cost of the wheeling is isolated, proportioned and the result used to adjust a bill rate, giving the analagous rate.

Example 20.3

Bill rate: concrete work

Concrete 1:3:6 in beds 100–150 mm thick, finished to receive quarry tile.	m^3	£98.93

Variation account

Concrete 1:2:4 in beds 100–150 mm thick, finished to receive quarry tile.	m^3

Labour will be the same for mixing and laying, so calculate material costs for both mixes as follows.

Basic cost of materials ascertained as:

Cement	£119.50 per tonne
Sand	£18.00 per tonne
Aggregate	£21.00 per tonne

Mix 1.3.6

	Vol.	*Mass/m³*	*Waste%*	*Cost/tonne*	
1 volume OPC	1	1.44	2.50	£119.50	£176.38
3 volumes sand	3	1.52	10.00	£18.00	£90.29
6 volumes aggregate	6	1.6	5.00	£21.00	£211.68

	hrs/tonne	hourly rate	
Unloading OPC: tonnes/hour	1.25	£6.50	£7.49
			£485.84
Shrinkage add 25%			£121.46
9 m³ costs			£607.30
1 m³ costs			£67.48

Mix 1.2.4

	Vol.	*Mass/m³*	*Waste%*	*Cost/tonne*	
1 volume OPC 1.44 t/m³ + 2.5% waste	1	1.44	2.50	£119.50	£176.38
3 volumes sand	2	1.52	10.00	£18.00	£60.19
6 volumes aggregate	4	1.6	5.00	£21.00	£141.12

	hrs/tonne	hourly rate	
1.25 hours/t unloading at £5.30/hour	1.25	£6.50	£7.49
			£385.18
Shrinkage add 25%			£96.30
6 m³ costs			£481.48
1 m³ costs			£80.25
Bill rate			£98.93
Profit and oncost is 20.00% so bill rate is equivalent to			120.00%
So net rate is			£82.44
Deduct cost of 1 m³ of material 1:3:6 mix			£67.48
If arguments earlier are correct, plant and labour cost is			£14.96
Add cost of 1 m³ of material 1:2:4 mix			£80.25
			£95.21
Add profit and oncost 20%			£19.04
Variation account rate per m³			£114.25

Example 20.4

Bill rate: facing brickwork

One brick thick wall built with facing bricks (PC £245 per 1000) in English bond, weather jointed in cement/lime mortar 1:2:9.	m²	£62.00

Variation account

One brick wall built with facing bricks costing £326.85 per 1000 in Flemish bond, weather jointed in cement/lime mortar 1:2:9.	m^2

Cost of bricks/thousand in variation account					£326.85
Cost of bricks/thousand in bill rate					£245.00
Extra cost of bricks per 1000					£81.85
English bond: bricks/m^2 faced one side					89
Flemish bond: bricks/m^2 faced one side					79
Additional material cost: £326.85 × (79/1000) – £245 × (89/1000) per m^2 is					£4.02
Bill rate					£62.00
Profit oncost 20%: so deduct one sixth					£10.33
There are 10 fewer facings and 10 more commons to lay in the Flemish bond wall so adjust labour costs:					£51.67
Labour: 2 bricklayers per hour at £8.00	2	8	£16.00		
1 labourer per hour at £6.50	1	6.5	£6.50		
Squad cost			£22.50		
Facings: output per tradesman 40 bricks/hour: 2 tradesmen					
Cost labour per 10 bricks:					
squad cost × 10/(2 × 40)	2	40	10	Ddt	£2.81
					£48.85
Commons: output per tradesman 60 bricks/hour: 2 tradesmen					
Cost labour per 10 bricks:					
squad cost × 10/(2 × 60)	2	60	10		£1.88
Add material cost					£4.02
					£54.75
Profit and oncost 20%					£10.95
Variation account rate per m^2					£65.69

The area of jointing and type of work remains the same for both the bill item and the variation item and the amount of mortar used will be the same.

This calculation shows that it is unwise to discount any factor in the assessment of an analogous rate. First impressions would make the QS believe that, because the principal material cost is over £80.00 per thousand dearer, the rate should be almost proportionally higher. In fact the saving on labour is such that it compensates for 25 per cent of the increase.

So if a client or member of the design team ever suggests that a more expensive facing brick is wanted on a contract, the QS might suggest a change to a bond which uses fewer facing bricks and thus lessen the impact of a large price hike in the budget.

Example 20.5

Bill rate: plaster work

2 cts Carlite plaster 13 mm thick to brick walls.	m^2	£7.56

Variation account

(a) 2 cts Carlite plaster 10 mm thick to block walls.	m
(b) 2 cts do on reveals not exceeding 300 mm wide.	m

Materials			
Carlite browning delivered site	*hours*	*rate*	£120.00
Unloading	1.25	£6.50	£5.20
			£125.20
Carlite finish delivered to site			£95.00
Unloading	1.25	£6.50	£5.20
			£100.20
	m^2/t	*th/mm*	
Browning covers at given thickness	140/160	11/8	
Finish covers at given thickness	450	1.6	
Labour			
Assume a squad of 2 craftsmen and 1 labourer	2	£8.00	£16.00
	1	£6.50	£6.50
Squad cost			£22.50

Materials cost/m^2:	
Browning 11 mm thick per m^2	£0.89
Browning 8 mm thick per m^2	£0.78
Materials saving per m^2	£0.11

	10 mm	*13 mm*
Output for squad for 2 ct work in m^2	5.5	4.5
For work in narrow widths the output is halved.	2.75	2.25

	General areas	*In narrow widths*
Cost to apply two coats browning 13 mm thick per m^2	£5.00	£10.00
Cost to apply two coats browning 10 mm thick per m^2	£4.09	£8.18
Labour saving per m^2	£0.91	£1.82

Variation account rate: item (a)

Bill rate per m^2		£7.56
Profit and oncost 20%: so deduct a sixth		£1.26
		£6.30
Deduct: material saving	£0.11	
labour saving	£0.91	£1.02
		£5.28
Add profit and oncost 20%		£1.06
Item (a) from variation account: rate per m^2		£6.33

Variation account rate: item (a)

Bill rate per m^2	£7.56
Profit and oncost 20%: so deduct a sixth	£1.26
	£6.30
Deduct labour cost for general areas 13 mm thick	£5.00
	£1.30
Add labour cost for narrow widths 13 mm thick	£10.00
	£11.30
Add profit and oncost 20%	£2.26
Item (a) from variation account: rate per m^2	£13.56
Cost per m^2	
Item (b) from variation account: rate per m	£4.07

Note that only the browning coat was adjusted and only the cost of the material. There is no need to bother with the skim coat of finish.

21
Information technology

Introduction

It can be generally stated that as soon as you take a computer out of its packaging and switch it on it is out of date! With this in mind it would be pointless to write a computing chapter describing the technology available at present, discussing the issues of size of computer memories, speed of modems, CD ROMs, etc. By the time this book is printed, published and sitting on your shelf, the technology will have advanced again, replacing the information and rendering it worthless.

To overcome this, this chapter will concentrate on the application of information technology (IT) systems and their appropriateness to construction, specifically relating to tendering. It is essential to be able to communicate with all the disciplines and levels within the construction profession. In order for IT to be a time saving and efficient tool, co-ordination of software and integration of information between construction professionals is the key to achieving an environment rich with innovation and opportunities for both clients and construction practices.

A multi-disciplinary framework of professionals could work together efficiently with the minimum of problems by the correct integration of IT, creating a 'one stop' service for clients from inception through to construction, use and decommissioning of a site.

IT applications are used extensively for a wide range of pre- and post-contract activities, and user knowledge of the applications is essential to achieve cross-disciplinary acceptance. Software applications must be user friendly, and construction professionals should need basic computer skills to be able to understand the package, not a degree in computer programming, i.e. you do not need to know how a car engine is built in order to drive a car!

This chapter will examine the use of IT in the construction profession under the following headings:

- Background
- Text/graphics
- Project management
- Data analysis
- Drawing/CAD
- Communications.

Background

The advances that have occurred in the field of computing and technology over the past few decades have been remarkable to say the least. The most noticeable difference is in the reduction in size of the equipment used today; it is more compact and portable while being significantly more powerful. Networking and communication mediums have also developed extensively – mobile phones, WAP and e-mail are used on a daily basis.

Computer systems can be mainly defined as a method of taking original data and interpolating it to produce an output. This technology can be developed further to include the hardware (the actual computer machine), the software (programs to perform the function, i.e. word processing, spreadsheets, etc.) and the data (text and figures) to provided overall a meaningful process.

The hardware provides the data processing power and this varies depending on the type of computer being operated. A personal computer has minimal power in comparison to a main frame computer system that is commonly utilized by banks, police, etc. to store large amounts of client/customer information.

The user face of application software technology has advanced greatly, 'user friendly' software can be operated without the need of advanced computer programming knowledge. The software package can be seen as a 'black box', the raw data is entered into the system by the user and an answer or solution is provided at the touch of a button. Input of the data is made simple by having blank spaces to be completed and step-by-step instructions, leading from one stage to another by a series of screens. Templates are provided and help windows are available to lead the user through the application.

It should be remembered that IT in any form is essentially a tool to aid the professional in completing their work – their function is not to replace the construction professional but to enable them to undertake their role more effectively providing an efficient service to the client. As mentioned previously some packages can be described as a 'black box', the information that is produced requires interpretation, provided by the experience and judgement of the professional, to maintain a service which the client can have confidence in.

Table 21.1 *Job function and software availability of certain building professions*

Professional	Job function	Software available
Architect	Involved with pre- and post-contract issues from sketch designs in response to client's brief at inception through to detailed working drawings and construction of the project.	Checking for compliance with brief, circulation analysis, housing type mix analysis, space needs provision, accommodation schedule production, circulation analysis, view analysis, plotting sketches, plans, elevations, section and perspectives, generation of schedules and specifications, plotting of working drawings. Building products data retrieval, automated detailing, structural member selection and sizing.
Engineer	Structural capacity of project, alternative options available to client and designer.	Preliminary structural computations, structural issues, calculations and problem solving, site investigation and analysis, drainage analysis, cut and fill analysis, accessibility analysis.
Project manager	Programming and planning of pre- and post-construction activities with respect to resource availability within time, budget and weather constraints.	Data programming and planning by activity data analysis, cluster and bubble diagram generation, overlay mapping analysis, site plan synthesis. Job costing, time and payroll functions, invoicing, Network analysis, precedence diagrams.

Continued

Table 21.1 *Continued*

Professional	Job function	Software available
Facilities manager	Controlling the running cost by maintaining the function of the building by providing schedule maintenance throughout the life of the building.	Maintenance and component reliability, programming and scheduling of maintenance, records of component life predictions for servicing requirements and failure rates.
Building services engineer	Providing the correct internal environment for the building function and the design of mechanical and electrical services.	Heat gain/loss calculations, insulation and shadow pattern analysis, natural and artificial lighting, sound transmission, reverberation time, mechanical and electrical services design, duct pipe, electrical network layout.
Quantity surveyor/ estimator	Involvement from initial conception to handover and running of building overseeing cost while providing value to the client.	Economic feasibility analysis, preliminary cost estimation. Generation and pricing of bill of quantities, cost analyses and project cost control. Pricing tender documents, bill rates, variations and final accounts.

Text/graphics

The construction industry needs to be able to communicate clearly and quickly to improve the efficiency and understanding of all the parties involved in the construction process. It will therefore come as no surprise when it is found that word processing is the computer software application most commonly used, whether to compile letters, reports, articles or other text documents.

Word processing has the main advantage of minimizing abortive typing, the editing features allow documents to be quickly drafted and edited, instantaneously checking spelling and grammar while providing an additional tool for formatting, allowing typeface, font size, and style to be set for each document. This would be impossible to achieve efficiently on a conventional typewriter.

Desktop publishing expands the powers of the basic word processing package further to include page layouts, photographs, graphics and diagrams to produce a published document of a professional standard.

Presentations are a major part of the construction industry, the manner in which a company presents itself could be the decision as to whether they win or lose a contract. Text and graphics can be excellent modes of communication in the form of a presentation either by the use of overheads, slides, a projection show or other means. Having the ability to present an idea, cost limit analyses, tender strategy, marketing tool or innovative technique can be accomplished by the presentation software that is available.

As previously mentioned word processing can save time on abortive work due to editing and corrections, it can also prove to be efficient by being able to produce standard documents, i.e. letters, reports, etc. Templates of documents are produced leaving spaces for individual pieces of information to be added. Mailing lists can also be created which means a standard letter can be sent out to a list of addressees with envelopes printed very quickly and easily. This can improve efficiency and the quality and standard of the finished document.

One of the main written documents produced regularly by the construction industry relates to specifications. NBS provide a software package, which can be subscribed to at various levels depending on user requirements. The system provides a library of specifications, guidelines and links to manufacturers' product information. The system is a concise and structured approach to specification writing.

Project management

An issue of great concern that clients, contractors and consultants have with regards to any project is time. An overrun can have a major effect on costs and the overall management and planning of the project. Meetings, appointments, long-, medium- and short-term achievements must all be monitored and recorded to ensure a project is co-ordinated in a structured manner.

The process or recording and monitoring by IT can take many forms – from palm held computers and electronic organizers, which can produce a task list of targets to be achieved and prompt windows to remind you of appointments, to specific software for scheduling of labour, plant and material.

From a construction focus, planning software systems to monitor progress on site from pre- to post-construction are commercially available. These are a useful tool for the management of any construction project either large or small to enable them to be able to identify resource requirements in advance and plan the consequent course of action required to fulfil the requirements minimizing time disruption due to unavailable labour, materials or plant. Being able to monitor this closely provides an indication of profitability, project resource planning and turnover.

Project managers must be experts in their field in order to prevent abortive or delayed work due to poor planning and lack of communication. The use of planning software can be a tool to aid the project manager to co-ordinate and monitor a specific project's progress, culminating in an overview of the whole process over a number of projects at varying stages of completion.

Data analysis

Data analysis can act as a guide with regards to future decisions on management strategy for the continuing development and expansion of a firm. By managing business information such as tenders received; tenders submitted; success of bids; design and production of materials, labour force and plant utilization, the contractor can subjectively provide services and products that are more specifically targeted to meet client demands and fall within the scope of an organization.

The use of spreadsheets within estimating has allowed information to be processed more quickly and accurately, repetitive calculations are minimized while efficiency and outputs are increased. Processing of data information and the advantages this creates for the user are similar to other software, such as word processing, etc. as it allows for editing of documents and undertakes the number crunching tasks with high speed and accuracy. Activities that could be undertaken by an estimator using spreadsheets include build-up of rates, reporting, resource processing and data analysis for future tendering.

Keeping up-to-date records of clients, contractors, builders merchants, hire companies and the respective information on each can be accomplished by a database. The information on each individual can be entered within certain defined fields and can be categorized either numerically or alphabetically. Once the information is entered it can be reproduced in various formats, i.e. relating to location, size, type of business, etc. This form of information handling can prove beneficial to the estimator when specific information requires to be sent to or retrieved from a select group of firms.

Activity databases are an extension of the above and provide an application of database software to aid in the selection and design of NHS hospitals based on the activities to be undertaken within the space. Information covering all activities undertaken within a hospital is correlated to the specific equipment, which would be required to provide the activities. The data is provided in the form of images to aid designers and specifiers with accurate briefing information which in turn provides estimators with detailed descriptions to aid with the accuracy of the tendering process.

Being able to evaluate by the prediction of probable outcomes of future complex issues and from the analysis of historical data provides the estimator with subjective guidance for the future planning and development of tenders and rates can be undertaken with data processing software. Numerical simulations, statistical analysis and multi criteria evaluation are techniques which can be utilized to help analyse and predict the performance of the firm in the past and how strategic changes will influence and impinge on resource allocations for the future.

Prediction of market trends influencing the growth of future investment and returns on money can be predicted with some degree of certainty by using formulae entered into a spreadsheet, for example the mortgage formula used in Chapter 3. The ability to calculate the predicted life and therefore the reliability and performance of a component or element can determine the design specification. For instance, will a high initial cost result in less maintenance and a longer useful life reducing running costs as opposed to a low initial cost resulting in increased maintenance and high running costs?

The term 'intelligent buildings' is commonly used to describe a building which enables occupants to undertake their function within surroundings that provide the correct physical and environmental atmosphere to encourage positive working and improve the performance of the business to be a leading market player in their field. This is accomplished by the use of IT to provide a building that is physically flexible to:

- accommodate the IT hardware and software requirements while allowing for expansion;
- provide the correct environment through heating, air conditioning, lighting, etc. monitored by thermostats, timers, sensors and detectors controlled using IT;
- be able to change the internal structure of the building to allow for expansion/downsizing of the core business activities; and
- manage the integration of the various technologies to provide for the three points above.

Drawings/CAD

Computer aided design (CAD) is probably the most commonly used specialist IT tool. It has been developed and upgraded through numerous versions and updates to provide a refined drawing tool that allows designers to produce 2D and 3D drawings of buildings. The drawing can be drawn to any scale and reproduced on paper to cover whole plans or detailed sections.

Drawing information can be stored and distributed electronically to surveyors, engineers and other construction consultants within the design team. The CAD software can be interactive to work with other software packages allowing quantities, specifications and planning to be compiled from the initial drawing software. This can be a very time efficient way of interpreting information and producing material relating to the specific project's working drawings.

Communications

Marketing can be made more effective by using the Internet to improve knowledge and customers. Customers can also be tracked easily and informed of new products and services on the Internet. Human resources and personnel functions rely increasingly on IT to manage information about staff, their availability and capabilities, and to make this information available across the business to other functions, for example the preparation of estimates/bids, job proposals, etc.

CD ROMs – manufacturers' catalogues and information on products.

Networks

The Internet or World Wide Web allows access to millions of web pages worldwide covering every subject area. This form of information can be useful to estimators to enable them to keep up to date with the latest legislation and market conditions. It can also inform organizations of what contracts are currently being tendered for and by whom. It is becoming increasingly important for construction companies and private practices to have their own web pages marketing their latest developments.

The Intranet allows employees within an organization to communicate with one another but does not allow access outwith the organization. All communication remains within the boundaries of the organization. This can be useful for confidentiality purposes.

E-mail is a method of communicating at high speed to and from anywhere in the world. E-mails can carry attachments, which allow large documents to be sent to anywhere in the world within minutes. This benefits the estimator as correspondence can be replied to quickly and easily prompting a quicker service with the most up-to-date information.

This chapter has provided an overview of information technology use in the construction industry. A useful web page to gain more detailed information is produced by Construction Best Practice. The web page address is http://www.itcbp.org.uk

22

Tender strategy

Introduction

A tender strategy is a plan made by a construction company in order to achieve growth, increased market share or other objectives as stated in the company plan. Construction companies are in business and therefore seek to gain wealth, which may be in cash or in assets such as buildings, land and plant.

It is generally perceived that in order to increase profit, turnover must also be increased. Conversely, a decrease in turnover will lead to a decrease in profits. Companies therefore generally plan for growth, with periods of stabilization in order to reinforce resources. The strategy may include the targeting of types of work or types of client, and may also involve marketing of the company by advertising or by making presentations to targeted clients.

Construction, however, is very different from manufacturing. If a manufacturer wishes to increase turnover, his plan will be based upon the development of a product, which is seen as being desired by the majority of the population. A car manufacturer, for instance, might develop a new model, which is fuel efficient, large, comfortable, powerful and available in many configurations. The costs will be calculated and the vehicle priced at or below the level of the competition. A particular market will be targeted, e.g. fleet owners, family motorists or the 20s age group. A massive advertising campaign will be aimed at the chosen market, with offers of test drives, free cars for a day, low finance and so on. Errors in estimating costs which result in low profits can be adjusted by price increases, or more likely through a mixture of model and specification revisions and price increases; for example, the 1.6i becomes the 1.6se, which is £1000 more but has air conditioning as standard.

A construction company wishing to increase turnover has basically to make the offer – the tender – attractive to the client and the client's professional advisers. The contractor's product is his construction service. The building has been designed by the architect; the client is interested in procuring the building as designed at the least cost. It follows, therefore, that the contractor wishing to increase turnover has to tender at a lower margin for profit than his competitors. Any profit on a fixed cost is profit; however, unlike the manufacturer, the contractor cannot adjust for estimating errors by price increases or revisions to the specification. The opportunity for a loss increases as the profit margin is reduced.

A construction company can operate in the short term without profit but cannot operate at all without cash. For this reason any plan for the company must include a cash flow forecast for the period of the plan. This forecast will be updated as tenders are won and costs are realized.

Within the company plan, and specifically within the compilation of the tender, account must be taken of the risk and uncertainty associated with particular projects.

It is the intention of this chapter to give a flavour of the above subjects, to allow an appreciation of the importance of accurate estimating in the company plan, and to discuss the strategies which can be devised in order to increase turnover and profitability.

The company plan

The construction company, unlike its manufacturing counterpart, can increase turnover without dramatically increasing its capital assets. A manufacturer looking to increase turnover by increasing the sales of a product must also increase production. This requires an increase in the production facility, usually a factory containing machinery owned by the manufacturer. This production facility requires capital. The construction company's production facility is the site, which is owned by the client. The construction company's manufacturing resource is construction plant, which is commonly hired.

A manufacturing company is a relatively stable environment with smooth cash flow; that is, income and costs are fairly static over a short period and flow in and out on a regular basis. It is rare for a manufacturer's liability to suppliers, etc. at any given time to exceed the assets. For this reason manufacturing companies can raise capital by selling shares, which basically represent the value of the assets of the company.

By comparison, a construction company is an unstable environment. Cash flows on individual projects vary; normally the highest expenditures and incomes fall in the middle third of a project. For all but the largest companies it is common for total liabilities to be greater than total assets. For these reasons, construction companies find difficulty in raising capital for projects by selling shares and are forced to finance projects through borrowing. On the other hand, construction company profits are potentially high in relation to capital invested.

The following example illustrates the application of a tender strategy. A builder, a private limited company, has a turnover of £10 million per annum. The builder's work comprises light industrial buildings, local health authority health centres and ambulance stations, local authority housing, petrol station forecourts and buildings, and small, high quality housing developments for a local speculator. All work is obtained by competitive tender except for the private housing work, for which the tender is negotiated.

In the company plan the builder has listed the following:

Assets

- Office block of 500 m^2 with yard and joinery workshop of 200 m^2. valued at £1 500 000.
- Sundry small plant and a wheeled tractor excavator with loader and backhoe. Total value £80 000.

Strengths

- Knowledge of the local area.
- Five highly skilled joiners and one apprentice.
- Four dedicated contracts managers.
- Capital invested of £400 000.

Weaknesses

- No track record with projects of over £4 million or with projects of longer than 18 months.
- No reputations with national clients other than one petrol company.
- Very traditional mode of operation.

Objectives

- To increase turnover to £10 million per annum during a five year period.
- To increase turnover to £3 million per annum in the following year.
- To sell top quality service rather than reducing tender margins.

Methods

- To target national clients, particularly hotel chains and supermarkets whose requirements are related to shopfitting and joinery.
- To investigate and sell quality control.
- To investigate and sell value management.
- To investigate and sell construction management and management contracting.
- To prepare sales literature, visit clients and advertise locally.

Resources

- Employ two contracts managers with experience in management contracting.
- Employ quality controller.
- Employ planner skilled in use of critical path method.
- Employ accountant to release director to new sales office.
- Employ one extra staff in sales, estimating/buying and accounts/wages.
- Build 300 m^2 extension to office.
- Reserve £40 000 for computers, fax and new telephone system.
- Negotiate bank overdraft facility of £2 million secured against office, workshop and yard.

Tender strategy

- Do not tender for design/build work at this juncture since this may prejudice commissions from local architects.
- Negotiate work on the basis of cost plus 5 per cent overheads and 4 per cent profit.
- In each month, tender for work on the basis of 5 per cent overheads and 2 per cent profit for the first £5 million, then 5 per cent overheads and 4 per cent profit for the remainder.
- If a tender is not accepted in first five attempts, reduce profit percentage addition to zero.
- Do not tender for work where the number of contractors tendering is greater than that recommended in the various codes of procedure for tendering.

In this example the builder is attempting to increase turnover through a variety of actions. To increase turnover in a construction market, which is buoyant and expanding, is easier than in a stable or a contracting market, since in the latter situation work is gained at the expense of competitors. To penetrate new markets or to increase market share within a sector requires extensive marketing and salesmanship.

It should be noted also that the plan contains a commitment to increase manpower resources, at current wage levels, worth approximately £700 000 per annum. This represents a 7 per cent overhead on the £10 million of increased turnover, and therefore in the first year of the new trading plan a small loss is anticipated on the overhead account. This will be corrected as turnover increases in the following year when no further expansion in staffing is anticipated. The company plan therefore is a set of interactive objectives aimed at achieving greater wealth.

In times of decreasing demand in the construction sector, the company plan will be a statement of the methods by which the contractor can retain resources in order to be in a good position to take on work when an upturn arises. This can result in the contractor planning to make a loss on selected projects, that is buying work.

Tender policy

The probability of winning

The probability of winning a tender is theoretically based on the level of mark-up for overheads and profit. If it is assumed that all estimators can accurately estimate the cost of carrying out work and that for a particular project all estimators plan for the same method of construction, then the only variable is the level of mark-up. It follows therefore that the winning tender will be that which has the lowest level of mark-up. This theoretical basis, however, does not perform well in a live situation since, as is discussed below, there are many factors which influence the final estimated figure.

One of the problems with competitive tendering is that the lowest tenderer generally wins. It is a fallacy therefore to think that by cutting margins on one estimate, subsequent estimates can be increased to cover the loss. Mark-ups for overheads and profit tend to move on an industry-wide basis, increasing when many projects are seeking contractors and decreasing when many contractors are seeking projects.

It follows that the successful contractor will be the one which observes the economic indicators and measures the likely future demand for work by reference to the overall performance of the economy. While this statement is generally true there is a problem in that the government, through its various agencies, is the construction industry's biggest customer – and the government tends to use the construction industry as a public spending regulator.

The construction industry therefore may receive little warning of a sudden downturn in demand. Good construction industry forecasters and planners will be watching the economic indicators in the same way as the treasury, and will attempt to guess at the manner in which projects from government agencies are likely to be released.

Reasons for variability between estimators' prices

As stated above, if it is assumed that all estimators can accurately estimate the cost of carrying out work, and that for a particular project all estimators plan for the same method of construction, then the only variable is the level of mark-up. This is plainly false. There are many reasons why estimators' prices vary, and the purpose of this section is to consider those reasons.

Items of responsibility

In pricing a tender, the estimator will have regard not only to the plans and bills of quantities but also to the specifications, preliminaries and conditions of contract. From these documents the estimator will build up a feel for the project and will overlay his base prices with an addition for complying with quality control provisions, work carried out under abnormal conditions, and so on.

Quantities

In situations where a bill of quantities is provided, the quantities are fixed. However, where the estimator is responsible for preparing quantities there is obviously scope for variability in measurement.

Material costs

Suppliers will offer special discounts to contractors who purchase large quantities of material in a given year. Thus large national contractors can obtain high levels of discount, which is reflected in the pricing of items or in the contractor's margin. Additionally estimators will use different allowances for off-loading, handling, storage and wastage.

Method of construction

A major factor in the variability of estimators' prices for items is the method for the carrying out of the work. The method will be derived at a meeting of the estimator, the planner and the buyer, together with the contracts manager and contracts quantity surveyor who have a preliminary designation to the project. At this meeting the method of construction, extent of subcontracting, timing of work items, availability of material, availability of in-house plant and other resources, and so on, will be discussed and will be recorded by the estimator. The pricing will be based upon the agreed method.

To illustrate the variability in method, consider the number of ways in which a 150 mm *in situ* reinforced concrete suspended slab can be constructed in accordance with given drawings and specification. The concrete, however it is delivered to the project, can be placed in a number of ways. First, it may be barrowed, lifted by hoist, and barrowed again into position. Second, it may be lifted by tower crane and the tower crane skip emptied in the required position. Finally, it may be pumped to the required position. The method chosen by the team will depend on the relative costs of each.

Remember that if a contractor has a tower crane in the plant yard, which is unlikely to be working otherwise during the concreting periods, then this could be the cheapest option, notwithstanding the fact that it may be the most expensive option on a hiring comparison.

Labour

Two variables are associated with labour, namely gross unit rate and productivity. Whereas the former can be relatively easily established (see Chapter 2), the productivity aspect is more difficult. Research into the productivity rates of labour has shown wide differences between projects and between estimators' assumptions. This can also be demonstrated by observing the differences between constants assumed in estimating books such as this.

Plant

As with labour, plant is subject to both cost and productivity variables.

Site conditions

While current standard contract conditions largely negate this subject as an unknown risk item, variations between cost and estimate can arise if the exact significance of the information provided is not appreciated by the estimator. For instance, if the presence of running sand in a trial pit is not noted by the estimator a low estimate may be made in the excavation work.

An example of a condition, which will affect the estimate but may not be contained in the contract documents, is a restriction in access. If the only direct route to the site has a height, width or weight restriction then this may affect the transport of plant and/or material.

Location

Construction costs vary throughout the UK. Studies demonstrate that building in the north of Scotland is characteristically over 20 per cent more expensive than building in the south-west of England. Local builders will be aware of the costs in their particular location, but national builders will adjust their standard data based on the location of the contract.

Escalation factors

In contracts which state that the tender price is to remain firm, estimators must make an allowance for increasing costs during the period of the project.

Contract time

It is common for the contract period to be specified by the design team, and the builder will plan resource requirements to fit the period stated. There is a theory that the method of construction adopted by the contractor is largely determined by the design and the period stated. That is, where there is a full design and a fixed period, all contractors tendering will choose the same method of construction. However, where parts of the design are left to the contractor, i.e. by including a performance specification in the contract documents, and/or the time is left to be stated by the contractor, then variability in work method between builders tendering is likely to occur.

Overheads and profit

As discussed above, these are the traditionally perceived areas of greatest variability. Profit is normally added to the tender amount as a percentage based upon a tender strategy. The overhead is a percentage added to the tender sum to cover the cost of the head office and supervision on site. An approximation of the percentage can be derived by taking the value of overheads from the previous year's audited accounts and dividing this sum by the turnover.

Contingency

This is an addition to cover contingent and unforeseen costs. It is accepted by the builder that the estimator will not have foreseen every eventuality and therefore a sum should be added based upon experience or on a calculation of risk factors (see later). However, since any addition reduces competitiveness, the contingency addition has to be carefully considered.

Cash flow and financing

The subject of cash flow is considered later, but it should be mentioned here that every project requires some degree of funding. The builder's income will not match his costs plus overheads and profit, since payment by the client is normally made up to six weeks in arrears and then less a retention of up to 5 per cent. The builder will be required to finance this shortfall in funding, normally through a bank arrangement. The tender should therefore include the finance charges on the overdraft.

Errors

It is commonly stated that the builder who wins a contract through competitive tendering is the one who makes the largest error. Procedures are outlined in the various codes of procedure for tendering for situations in which mathematical errors occur. If the contractor discovers an error of any other nature, the only recourse is to withdraw the tender (the offer) before it is accepted by the client.

Cash flow

The cash flow of a builder is defined as the resultant of income and expenditure expressed as a function of time. Cash flow is considered positive when income exceeds expenditure and negative when expenditure exceeds income. The profile of a typical project (Figure 22.1) shows a negative cash flow for the first 12 months of an 18 month contract, i.e. the majority of the contract period.

The expenditure is normally shown as a curve, since costs arise on a weekly basis for labour and a monthly basis for subcontractors, although for logistical reasons the valuation and payment dates are spread through the month. Suppliers are normally paid at the end of the month following the month in which the material was delivered.

Income, on the other hand, is by interim payment, which is made by the client on a monthly basis, normally two weeks following the end of the valuation month. This is represented on the diagram by the vertical bars.

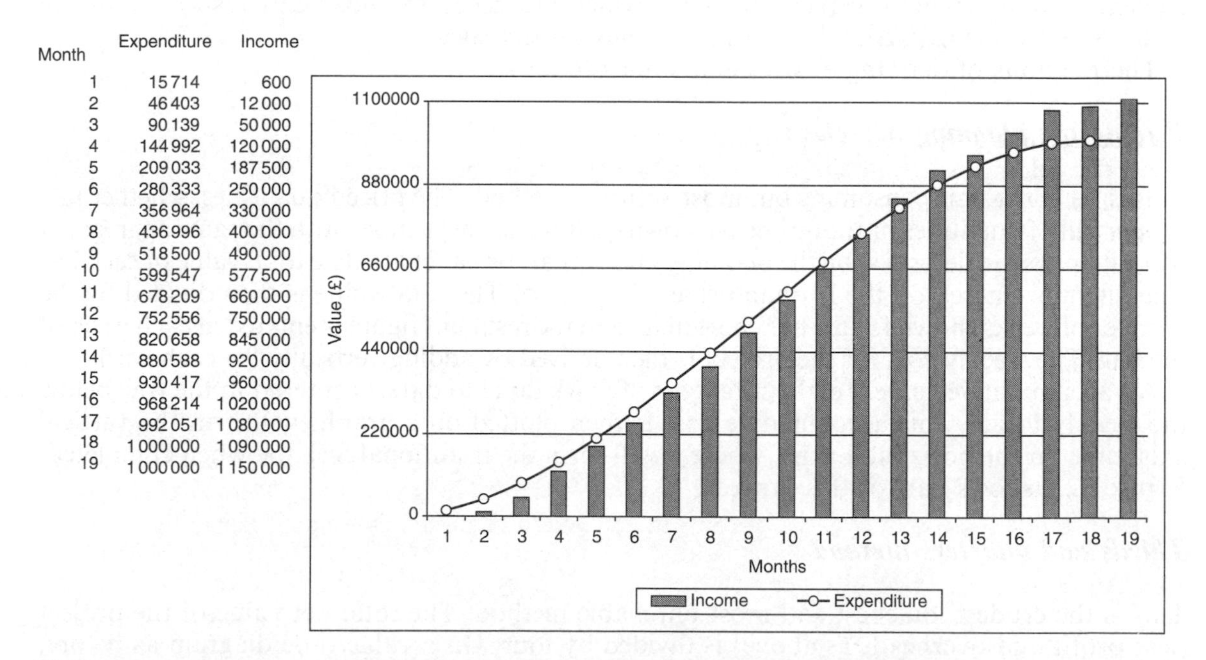

Month	Expenditure	Income
1	15714	600
2	46403	12000
3	90139	50000
4	144992	120000
5	209033	187500
6	280333	250000
7	356964	330000
8	436996	400000
9	518500	490000
10	599547	577500
11	678209	660000
12	752556	750000
13	820658	845000
14	880588	920000
15	930417	960000
16	968214	1020000
17	992051	1080000
18	1000000	1090000
19	1000000	1150000

Figure 22.1 *Cash flow diagram: the curve represents expenditure, the bars income*

As stated earlier in this chapter, the highly volatile construction market, the financial structure of builders and the irregular peaking of demands for finance make building companies very vulnerable to temporary shortages of cash. The problem facing companies is therefore one of convincing the financier, usually a bank, of the fact that the shortage is only temporary. A bank would prefer to see a cash flow forecast of the type illustrated for each project, and an aggregated forecast for the company as a whole.

Since large national contractors have many projects at various stages of completion, their aggregated cash flow tends to a horizontal line, that is a constant demand for cash. Small and medium companies with a smaller number of projects tend to show wide fluctuations in the company demand for cash.

This demonstrates the paradox that large national companies who have the resources to carry out a cash flow forecast have no need to make one, whereas small and medium companies have fewer resources but need to monitor cash flow very closely.

Construction of a cash flow diagram

A cash flow diagram for a tendered project is important, since it permits the amount of finance required for funding the project to be determined. At the tender stage it also allows the repricing of items to minimize the finance required. A cash flow forecast, as illustrated in Figure 22.1, may be constructed as follows.

Derive cost curve

There are four methods of deriving a cost curve, which are described below. In each method the cost curve represents the tender value of the project excluding head office overheads and profit.

Head office overheads are deducted from the project cash flow and added to the aggregate of all project cash flows, which forms the company cash flow. Overheads are normally a constant value per month; they are costs, which are generally incurred irrespective of the value of work being carried out on the various project sites.

The methods of deriving a cost curve are as follows.

Pricing the planning bar chart

This is the most time consuming but most accurate method. The priced quantities, whether in a proper bill of quantities or the estimator's own quantities, are grouped into the same aggregated work items as are described on the planning Gantt chart or bar chart. The total value of each bar chart item is entered on the left-hand side of the chart. The total value is then divided by the number of weeks shown for the bar chart item, and the resultant figure is entered in each week of the bar. The weekly cost for the project is then derived by adding vertically for each week.

An accumulative value, i.e. the total cost of work done to date, is entered in the row below the weekly totals. This accumulative cost is then plotted on a graph (value on the vertical axis, time on the horizontal axis), where it will form the traditional S shape when completed. Figure 22.2 shows part of this process.

Thirds and quarters method

This is the crudest, quickest and most unreliable method. The total net value of the project (less profit and overheads as above) is divided by four. On a value/time diagram as before, four points are marked. The first is at (0,0). The second is at one-third contract time and

Pricing the planning chart

	Value	Week 1	Week 2	Week 3	Week 4	Week 5
Set up site	4000	2000	2000			
Piling	20 000		5000	5000	5000	5000
Excavate trenches	16 000			4000	4000	4000
Concrete foundations	24 000				6000	6000
Brick substructure	38 000					7000
Subtotal		2000	7000	9000	15 000	22 000
Accumulative		2000	9000	18 000	33 000	55 000

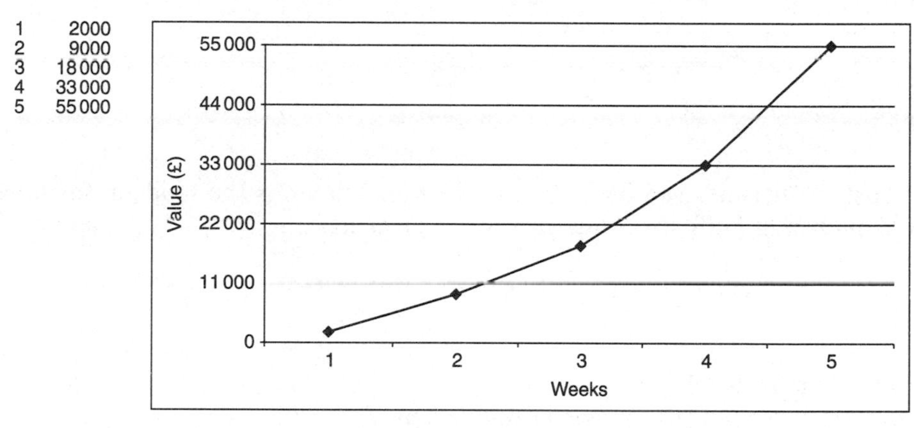

Figure 22.2 *Priced bar chart and S curve*

one-quarter project net value. The third is at two-thirds contract time and three-quarters project net value. The final point is at full contract time and project net value.

Figure 22.3 shows this method for the example of Figure 22.1, that is for a project total value of £1 100 000, a project net value of £1 000 000, and a contract period of 18 months.

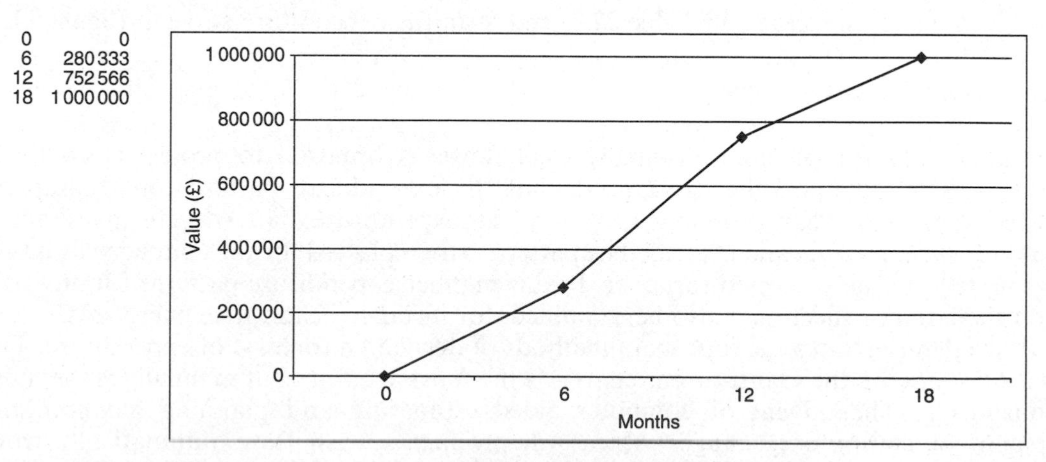

Figure 22.3 *The S curve by the thirds and quarters method*

Table 22.1 *Parameters in Hudson formula*

Contract value	Parameters C	K
£10 000 to £30 000	−0.409	7.018
£30 000 to £75 000	−0.360	5.000
£75 000 to £120 000	−0.240	4.932
£120 000 to £300 000	−0.200	4.058
£0.3 m to £1.2 m	−0.074	3.200
£1.2 m to £2.0 m	0.010	4.000
£2.0 m to £3.0 m	0.110	3.980
£3.0 m to £4.0 m	0.159	3.780

Formula method

In this method the cost curve is derived by formula. The best known is the Hudson formula, developed for use in planning DSS spending on capital projects:

$$y = S\left[x + Cx^2 - Cx - 1/K(6x^3 - 9x^2 + 3x)\right]$$

where:
y is the cumulative value of work;
x is the month in which expenditure y occurs divided by P;
P is the contract period in months;
S is the contract value; and
C and K are parameters based upon the contract value (see Table 22.1).
Using the same example as above, 18 cumulative values are derived:

Project total value	£1 100 000
Project net value	£1 000 000
Contract period	18 months

The derived values are shown in Table 22.2, and form the expenditure curve in Figure 22.1.

Balance sheet methods

More usually used for planning company cash flows as opposed to project cash flows, this method is based upon the assumption that the expenditure pattern for a company is repeated year after year. For any company the expenditure patterns are investigated from balance sheet records and plotted. The pattern is then plotted for the coming year based proportionately on the projected turnover. In this manner expenditure patterns for any cost centre on the balance sheet may also be examined, for instance head office overheads.

The above demonstrates the four main methods of deriving a forecast of expenditure. The first method – pricing the Gantt or bar chart – is the most accurate but manually is the most time consuming. The advent of computer aided estimating and planning has seen the development of computer packages which will produce a cash flow automatically from estimating and planning data. This is described in Chapter 21.

Table 22.2

Month	Value of y (£)
1	15 714
2	46 403
3	90 139
4	144 992
5	209 033
6	280 333
7	356 964
8	436 996
9	518 500
10	599 547
11	678 209
12	752 556
13	820 658
14	880 588
15	930 417
16	968 214
17	992 051
18	1 000 000

Insert income bars

Having derived the cost curve, the forecaster may now insert the income bars. The method involves first marking the cost curve at monthly intervals and reading off the monthly expenditure values. To these expenditure values is added the percentage value of overheads and profit, since the income received from the client will include these percentage additions. Payments from the client are normally received 14 days following the month in which the work was carried out, and therefore the income is plotted at the midpoint of each month as shown by the bars in Figure 22.1.

Review cash flow

The final operation in the cash flow analysis at tender stage is to review the financial implications of the forecast in terms of determining the amount of money which is required to fund the contract. As stated earlier, this is normally obtained from a bank and will be more easily acquired if it is supported by a forecast.

The review stage permits the uneven allocation of the overheads and profit fund to minimize financing, therefore lowering the interest payable and reducing the tender. This operation is commonly called 'front loading' the tender. In extreme cases it is possible to artificially increase the amount in the overheads and profit fund by making negative overheads and profit additions to items of work which will be carried out later in the project. An example of a review to minimize financing is demonstrated in Figure 22.4 using the values obtained in the formula method example above.

Financial manipulation may also be used to artificially force a contract into loss in the earlier period where there is a requirement by the company to reduce trading profit in any

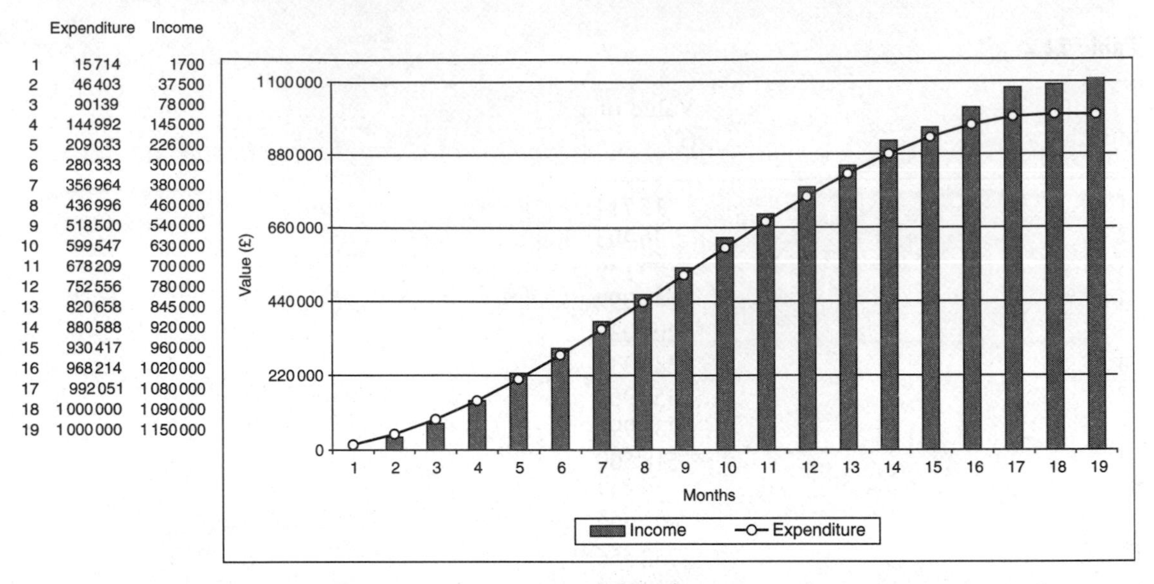

	Expenditure	Income
1	15714	1700
2	46403	37500
3	90139	78000
4	144992	145000
5	209033	226000
6	280333	300000
7	356964	380000
8	436996	460000
9	518500	540000
10	599547	630000
11	678209	700000
12	752556	780000
13	820658	845000
14	880588	920000
15	930417	960000
16	968214	1020000
17	992051	1080000
18	1000000	1090000
19	1000000	1150000

Figure 22.4 *The result of financial manipulation: the curve represents expenditure, the bars income*

particular trading year. For instance, if the current trading year is considered to be very profitable then tenders for projects in the remainder of the year can be 'back loaded'. That is, the overheads and profit will be attributed to those items carried out near the end of the contract, thereby transferring money from one trading year to the next. This has an advantage in that less corporation tax may be payable, but carries the risk that if the project is not completed for whatever reason then the builder will incur a loss. It also means that the money transferred is effectively lying in the client's bank account earning interest instead of in the builder's account. However, if the contract is one with a formula price adjustment clause then the loss of interest will be offset by a high increase in price recovery.

Risk and uncertainty

In a situation of risk, the result of an event is not known precisely but it is possible to determine:

- The number of possible results
- The probability of each possible result
- The value of the outcome of each possible result.

Uncertainty is defined as being a situation in which the determination of these factors is not possible. It is difficult therefore to calculate the effect of uncertainty upon the tender, and a discussion of uncertainty is out of place here.

A builder will see risks in two distinct categories:

- Those that can be insured against
- Those that cannot be insured against.

As far as the tender is concerned, risks such as fire, vandalism and theft will be insured against and the premium will be added to the tender in the preliminaries section. For those risks which may not be insured against, such as strike action, weather and failure in the power supply, an assessment will need to be made of the three factors above. Generally a good checklist of risks is given in the standard forms of contract, being those items for which the contractor is entitled to an extension of contract time but is prevented from claiming any loss and/or expense. In this situation the value of the risk item is the cost of maintaining the site with no work in progress.

The risk cost can be expressed as:

$$\text{risk cost} = \text{probability of event occurring} \times \text{period of event} \times \frac{\text{cost of delay}}{\text{unit of period}}$$

For example, if there is a 30 per cent probability that the site will be held up for one week due to bad weather and the cost of maintaining the site for one week (including non-productive payments) is £10 000.

$$\text{risk cost} = £0.30 \times 1 \text{ week} \times £10\,000 \text{ per week} = £3000$$

The £3000 will be added to the contract either in the preliminaries section or to the profit and overheads fund.

Some builders will have a checklist of risk items to analyse, although it is generally perceived that little work has been done in the direction of creating a database for effective risk management.

Appendix A

Cost of mortars, renders and fine aggregate screeds

Introduction

Plastic mixtures of cement and fine aggregates, or of cement, lime and fine aggregates, are used for a variety of purposes in the construction industry. The most important of these are:

- As mortars when building bricks and blocks.
- As coatings on walls, ceilings and soffits, i.e. roughcasts and renders on a variety of bases such as brick, block, concrete, stone and lath.
- As coatings on subfloors, i.e. screeds to receive floor finishes.

This appendix deals with the calculations necessary to provide the cost of one cubic metre of mortar, render or screed ready for use on site. The term 'mortar' will be used for the remainder of the appendix.

The sample calculations have all been done using a computer and a spreadsheet. The computer is an IBM compatible PC and the spreadsheet program used is called Excel.

By adjusting basic costs and hours on a spreadsheet it is possible to obtain a thorough appreciation of the effects on the final cost of the material produced. Moreover, this can be done quickly, it can be updated as new raw materials and labour costs arise, and the results can be preserved for future use either as a computer file or on hard copy.

Data for mortar costing

The basic data required for mortar cost calculation are as follows:

Material costs and density constants

Materials are generally priced per tonne delivered to site. Constants of bulk density for aggregates vary according to moisture content and the mass of the aggregate material.

Shrinkage

The shrinkage of the total volume of dry materials to the final bulk of the wet mix is usually expressed as a percentage. It can vary depending on the type of aggregate used, the length of time for which mixing takes place, the addition of air entrainers, etc. and the quantity of mixing water added.

Cost of labour

Small concrete mixers may require only one labourer to allow efficient operation. Larger machines require not only additional men but also some mechanical plant to load the drum efficiently. This appendix will confine itself to examples requiring up to two men, loading with hand tools.

Cost of plant

Most contractors own their own concrete mixer(s). However, use of these or hired equipment does represent a charge against a contract. How this charge is priced out is shown in Chapter 3. The hire rates used in this appendix have been extracted from trade sources current at the time of calculation. They represent only the 'capital' element of the cost of the machine; fuel, oil, grease, cleaning, etc. are added as part of the running cost; i.e. the mixing operation. Maintenance of owned plant is treated as an overhead and is thus included in the oncost added to the final bill rate. Hired plant is maintained by the hire company, and this cost is reflected in the hire rate. Delivery to and uplift from site have not been included in the calculations.

Operational hours

While plant is charged to the site for each hour it is physically present on site, it does not always follow that it is actually productive for each of these hours. Generally, hire of concrete mixers is by the week. The drum seldom turns for 39 hours in any normal week. From his knowledge of the quantity of mortar required and the number of tradesmen applying it, the estimator will calculate how long the mixer must operate to provide that quantity. For example – and a fairly simplistic one – suppose the estimator has a bill with one measured item: a one brick thick wall. The quantity is 2000 m^2, and he proposes to put a squad with eight tradesmen onto the bricklaying. Eight men will lay 400 to 500 bricks per hour, so the number of hours is $(2000 \times 118)/500 = 472$, or about 12 weeks. The brickwork requires $[(2000 \times 118)/1000] \times 0.6 = 142\,m^3$ of mortar.

Using a 5/3.5 concrete mixer with an output of 3.5 ft^3 (0.102 m^3) of wet mortar per mix, the drum has to be loaded and emptied 1390 times in that 12 weeks. The output of a 5/3.5 concrete mixer is generally reckoned to be 1 m^3/hour, so 142 m^3 will take 142 hours. This is roughly 6 minutes per batch. A more optimistic output might be 5 minutes per batch of 0.102 m^3, giving a running time of 116 hours. Divide by 12 weeks and we have the number of 'operational hours' per week for the mixer – say 10 hours. So the hire cost is divided by 10 hours in this case when calculating the cost of mixing.

Of course, the bricklayers might well break all records and finish the work in 9 weeks. In that case the mixer would be operational for 13 hours per week. Hire costs are saved, there is no saving in fuel or other consumables, and the bricklayers would want a good bonus or piecework rate.

If the reader has followed the argument so far, it will be obvious that a larger mixer would not have to run for so long – but it would probably make more mortar than the bricklayers could comfortably use before it took an initial set. Then again, a smaller mixer would run for longer and would produce smaller batches – but these might be so small that the bricklayers would keep running short of mortar, and that would hurt their bonus earning possibilities. So the estimator has to use not just cost estimating but also resource scheduling, so that plant is matched to the job and the labour force available.

Generally, hours running are more than this example shows. A figure of 20 to 35 hours per week is reasonable. Not that the machine is then producing just mortar for building; it may well be providing small quantities of concrete, screeding and rendering material.

Manual costing

A typical calculation would look as follows.

Cement mortar, 1:3, machine mixed in a 5/3½ mixer

Materials		
Ordinary Portland cement delivered site per tonne		£119.50
Labour unloading: 1.25 tonnes/hour at £6.50/hour		£5.20
		£124.70
1 m³ OPC weighs 1.44 tonnes at £124.70/tonne		£179.57
3 m³ washed building sand weighs 4.56 tonnes at £19.50/tonne		£88.92
		£393.19
Shrinkage on mixing is 20%: so add 20/(100 − 20) = 25%	20.00%	£78.64
Cost for 4 m³ of material		£471.83
Therefore 1 m³ of material costs		£117.96
Labour and plant		
5/3½ mixer hire rate per week	£27.00	
1 labourer in charge at £6.85/hour	£267.15	
1 labourer in attendance at £6.50	£253.50	
Cost per week	£547.65	
Machine is used for 30 hours per week so machine cost per hour is:	£14.04	
Cleaning and maintaining 1/2 hour per 8 hour day at £6.50/hour	£0.41	
Fuel: petrol 1.14 litres/hour at £0.80 litre	£0.91	
Oil and grease 10% of fuel cost	£0.09	
	£15.45	
Output mixing mortar is 1.15 m³ per hour so cost to mix 1 m³ is:		£13.44
Cost of 1 m³ of mortar without profit or oncost		£131.39

This is a fairly simple, if time consuming, calculation if different mortar types have to be costed or different mixers and operational hours are to be tested.

Spreadsheet costing

A quick way to do this is to set up the calculation on a computer. There is no need to have purpose written software, and no need to be able to program. The costing uses a standard piece of office software called a spreadsheet.

The computer printouts shown in the following are fairly self-explanatory. The initial blank spreadsheet or template was made up for use by students, and incorporates the following principles:

1. Information is laid out in screenfuls or worksheets (see later), i.e. discrete operations are confined to the data that can be displayed on one screen at a time.
2. The important parts of the sheet – the answer you want and the supporting statements – are on the first screen.
3. Principle 2 means that printing out the first screen is all that is required to keep a hard copy for the file.
4. Details of the calculations are put on later screens.

The examples which follow are all based on 30 operational hours per week for the mixer.

Example 1

Up-to-date versions of the spreadsheet Excel hold data and calculations in 'Workbooks'. A workbook is saved as a file under a single name. Each workbook can be a single screen of data and calculations or be made up of a number of such screens. Individual screens are called worksheets and within the workbook file, these can be named. Figure A.1a shows the first screen (a worksheet called 'Mortar') for the costing of a cement/lime mortar 1:2:9 mixed using a 5/3.5 diesel fuelled concrete mixer without the addition of plasticizer or anti-freeze. The other two worksheets (screens) are named Materials and Plant and Labour and are utilized for calculations and the storing of constants, costs, etc. The reason for keeping this information in separate worksheets is because it would be updated once for the job in hand and the various mortars, renders or screeds costed out on the Mortar worksheet before updating the materials worksheet for the next job.

Figure A.1a is a printout of the Mortar worksheet where the workbook has been set up for costing out a 1:4 mortar mix without any additives. Figure A.1b is a printout of the Materials worksheet showing the materials costs used for the mortar mix and Figure A.1c is the Plant and Labour worksheet showing the constants and costs used for that mix.

A detailed explanation of the sheets could be given but it is assumed that students and practitioners have a working knowledge of spreadsheets and have access to the Internet in which case all the spreadsheets used in the book are available on Butterworth-Heinemann's own web site (see note at front of book). Nothing beats getting the spreadsheet downloaded and playing around with it!

Before altering any data on the spreadsheets, may we suggest that the reader attempts to price out the m^3 cost of a variety of mixes. See what happens if you choose two mixers for a given mix and look at the effect of choosing a plasticizer with a lime mix. Each of these events will trigger an 'Error message'.

Now try updating the costs of materials, hire costs and labour rates and see the effect that these updates have on the m^3 costings.

Finally, get into the shaded cells and look at the formulae embedded in the spreadsheet and then try extending the spreadsheet to incorporate, say, another mixer size – a $3/2\frac{1}{2}$. And then perhaps another additive, say and air entrainer or a water proofer.

Of course, should you uncover any errors in the spreadsheets, let us know through the publishers.

MORTAR - ANY MIX - ANY PURPOSE **JOB:** *Housing, Anytown*

Enter proportions for cement, lime and sand in boxes below headings Do NOT alter data in shaded boxes
Choose a mixer by entering a '1' under Petrol OR Diesel. Only choose ONE mixer.
Choose a plasticizer or anti-freeze, by entering a '1' in the box below each heading
Additives are calculated from the proportion of the cement by mass

Error messages

OP Cement	Lime	Sand	Anti-freeze	Plasticizer
1	2	9		

Mortar cost £95.91 m^3

The default shrinkage is 25 % or give an alternative
shrinkage allowance here %
The shrinkage allowance being used is 25 %

Choose a mixer size and fuel type by entering a '1' in the box opposite any **one**

Mixer size	Hire cost		Petrol	Diesel	Output in m^3	
4/3	£24.00	per week			0.95	per hour
5/3½	£27.00	per week		1	1.15	per hour
7/5	£35.00	per week			1.75	per hour

To update the cost of materials, labour or plant hire click on the appropriate worksheet title below

The workbook is set to print only this worksheet as your record of the mix calculated and will print landscape on A4 paper.

Figure A.1a *Printout of Mortar Worksheet*

Materials costs Last updated 10-06-2002 Do NOT alter data in shaded boxes

Material	Cost delivered site	Mass/m^3	Unload 1 tonne - hrs	Parts by Volume		Calculation of Materials costs
OPC	£119.50 tonne	1.44	1.25	1		£188.93
Builders lime	£210.00 tonne	0.6	1.25	2		£257.85
Washed building sand	£19.50 tonne	1.52	–	9		£266.76
		Application rate				
Anti-freeze	£1.10 litre	80	1/tonne	0		£0.00
Plasticizer	£1.95 litre	4	1/tonne	0		£0.00
						£713.54
			Shrinkage – add %		25	
			So use %		33.33%	£237.85
						£951.38
			Volume of agg. + shrinkage		12	
				Cost per m^3		£79.28

If shrinkage on mixing is say 20% then original volume is 80% so addition at this point is 20/100−20 = 25%

Figure A.1b *Printout of Materials Worksheet*

Plant and Labour Costs

Last updated: 10-06-2002

Labour:

	Rate			Costs
Labourer in attendance	£6.50	1	39	£253.50
Labourer in charge	£6.85	1	39	£267.15
Clean mixer 1/2 hr per day for 5 days				£16.25

Plant with consumption and output rates

Mixer	Hire cost/ week	petrol per hour litres	diesel per hour litres	Lub oil per hour litres	grease per hour kg.	petrol per hour	diesel per hour	lub oil per hour	grease per hour	Mixer size	Weekly hire cost of chosen mixer		Weekly fuel, oil and grease cost
4/3	£24.00	0.24	0.20	0.03	0.003	0.000	0.000	0.000	0.000	0	£0.00	£0.00	£0.000
5/3½	£27.00	0.40	0.29	0.04	0.004	0.000	0.290	0.040	0.004	1	£27.00	£36.72	£9.720
7/5	£35.00	0.55	0.47	0.05	0.005	0.000	0.000	0.000	0.000	0	£0.00	£0.00	£0.000

Operational hours per week: 30

Weekly cost: £573.62

Fuel Etc.

Petrol	£0.80	litre
Diesel	£0.80	litre
Lube oil	£2.00	litre
Grease	£3.00	kg.

Hourly cost: £19.12

Output per hour: 1.15

Cost/m^3: £16.63

Figure A.1c *Printout of Plant and Labour Worksheet*

Index